GUIDE TO SUCCESSFUL MARKETING AUTOMATION

THE ULTIMATE NO HOLDS BARRED GUIDE TO TECHNOLOGY, AUTOMATION, & ARTIFICIAL INTELLIGENCE

Dan S. Kennedy and Parthiv Shah

Ep
Entrepreneur PRESS®

Entrepreneur Press, Publisher
Cover Design: Andrew Welyczko
Production and Composition: Alan Barnett Design

Library of Congress Cataloging-in-Publication Data

Names: Kennedy, Dan S., 1954- author. | Shah, Parthiv (Businessman), author.
Title: No B.S. guide to successful marketing automation : the ultimate no holds barred guide to technology, automation, & artificial intelligence / Dan S. Kennedy and Parthiv Shah.
Other titles: No bullshit guide to successful marketing automation
Description: Santa Ana, CA : Entrepreneur Press, [2024] | Series: No B.S. | Includes index. | Summary: "Dan Kennedy and Parthiv Shah leverage their extensive marketing expertise to teach you proven, no-nonsense strategies for achieving successful marketing automation"— Provided by publisher.
Identifiers: LCCN 2024005386 (print) | LCCN 2024005387 (ebook) | ISBN 9781642011708 (paperback) | ISBN 9781613084823 (epub)
Subjects: LCSH: Marketing—Technological innovations. | Automation. | Artificial intelligence.
Classification: LCC HF5415 .K54155 2024 (print) | LCC HF5415 (ebook) | DDC 658.800285--dc23/eng/20240524
LC record available at https://lccn.loc.gov/2024005386
LC ebook record available at https://lccn.loc.gov/2024005387

CONTENTS

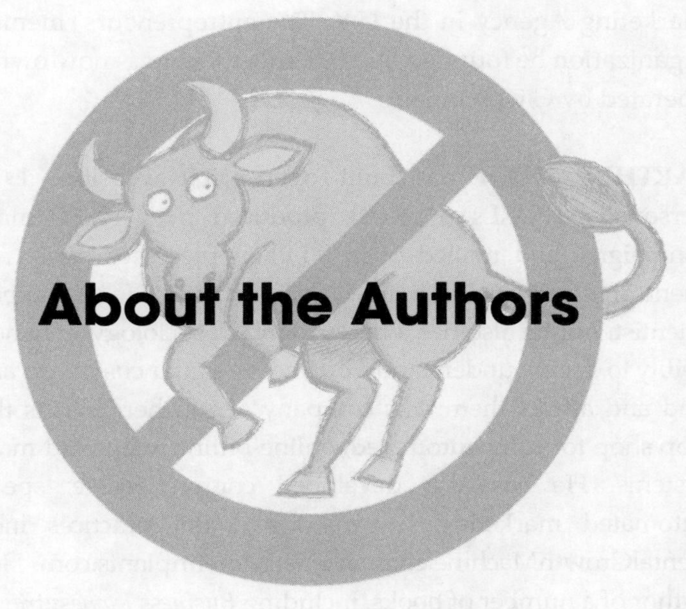

About the Authors

DAN S. KENNEDY is a from-scratch multi-millionaire serial entrepreneur, and a sought-after direct marketing strategy consultant and direct-response copywriter who has frequently, intimately worked with start-ups and growth companies in creating multi-million- to billion-dollar successes. He influences well over a million entrepreneurs, small business owners, private practice professionals, and sales professionals each year through the ever-growing NO B.S. book series, the NO B.S. Inner Circle/ Magnetic Marketing newsletter, courses, events and coaching programs, and other newsletters and speaking engagements. While famously a modern Luddite personally, for his clients he utilizes all available media and online opportunities and, as of this writing, is a shareholder in numerous marketing automation and tech companies including Keap (Infusionsoft), Snowflake,

HubSpot, Salesforce, and, of course, Microsoft, Apple, and Amazon. Dan is also on the advisory board of The Internet Marketing Agency in the U.K. The entrepreneurs' membership organization he founded, No B.S. Inner Circle, is now owned and operated by ClickFunnels.

PARTHIV SHAH has built over 1,000 websites, 1 million personalized URLs (PURLS), produced over 10,000 marketing campaigns, and mailed over 1 billion pieces of direct mail for clients in diverse fields, in the U.S. and abroad. He is an expert data scientist, but he also has a background in sociology, enhancing his ability to deeply understand a business's avatar customers and then find and attract them. His company eLaunchers.com is the one-stop shop for fully automated, online-offline integrated marketing systems. He has also developed comprehensive, specialized automated marketing systems for dental practices including DentalGrowthMachine.com and SellMoreImplants.com. He is the author of a number of books, including *Business Kamasutra*.

About Contributing Authors & Experts

Preston Bates has appeared on the *CBS Nightly News, CNBC, Bloomberg* and *PBS,* and in *Forbes, The Wall Street Journal, The Economist, The Daily Beast, Politico,* and *The New York Times*. He has worked with visionary leaders in marketing automation and been #2 at the #1 software company in the field. Follow him on Instagram @prestonsbates.

Dr. Carlo Biasucci, DDS, is the founder of Elite Practice, Canada's largest provider of training for dental office teams. He is the author of four books and numerous resources for dentists. www.TheElitePractice.com

Dipiti Kala is a retired squadron leader in the Indian Air Force and is founder of Tarika Consultations in India. Over 500 consultants and coaches have benefited from working with her. As an aeronautical engineer, her interest is in experimenting with A.I. in various business applications. www.diptikala.com

Seth Greene is the originator of the 50-DREAM-VENTURE-PARTNERS SYSTEM, for having a business promoted every week for a year. He is the only person to be nominated three years in a row for the Kennedy Inner Circle (GKIC) Marketer of the Year Award. He has been endorsed by Amanda Holmes, Jay Abraham, Rich Shefren and Russell Brunson. www.marketdominationprogram.com

Jeff Hunter is the creator of the A.I. PERSONA METHOD and has consulted with over 100 companies on its use. He is a highly successful direct-response copywriter and one of the very few to get A.I. to write or assist with writing sales copy that rivals the output of top pros. www.jeffhunter.com

Acknowledgements

Terri Angela Lopez and Everte Farnell were vital contributors, taking my thoughts in their raw form and organizing and articulating them in a reader friendly way. They are both excellent freelance writers and editors.

Marketing Automation Companies, Platforms, and Software You Will Encounter in This Book

(That You May Want to Research after Reading This Book)

- Constant Contact
- ClickFunnels
- HubSpot
- Kaep, formerly Infusionsoft
- Mailchimp
- Salesforce

Important Notice

Marketing Automation is a subject that defies doing a book about it. Although the success principles and strategies for its use hold steady, many, many details change like the weather. For that reason, for at least three years from this book's copyright date, its Authors will maintain a FREE UP-DATE SERVICE. Up-Dates of various kinds will be posted at and/or available at:

eLaunchers.com/tech-updates

Introduction

How to Make Your Business Much More POWERFUL

by Parthiv Shah

I want to congratulate you on picking up this book, and I want to make you a promise… This book WILL help you build, automate, and use profit-producing systems in your business. We discuss how technology can help you do that, and we share different tools that you can use. But this is a marketing book, NOT a technology book. What I mean is that we discuss the use of technology to an end. For a purpose. More sales, more profit, and a higher valuation of your business. More power to accomplish more of your vision and ambition, faster. The ability to do things that now flummox you. This is what technology, properly organized and used, can do for you—like nothing else can. However, just because a Pandora's box of tech exists does *not* mean you should feel pressured to use it. As you'll see, I use many different companies' tools, but different ones for different clients and their different purposes.

We are going to discuss software that will help you generate more and more profitable sales, on autopilot, with less money invested in getting each sale. By the way, the fact that we will most likely be able to bring down your Cost Per Sale (we discuss this and several metrics later in the book) is not an excuse to spend less on marketing. Instead, our goal is to empower you to confidently invest more, to get more growth, and more speed of growth. A key goal of Marketing Automation is *leverage*. This comes from power. The power to do more, do it faster, do it better, do it more consistently, and do it more cost-efficiently so that growth or expansion can be friction-less, and you can exert competitive advantage.

To gain these benefits, you need to have your entire marketing as a fully integrated SYSTEM, not day-by-day, reactionary, random acts. My co-author, Dan Kennedy, says that the more random you are, the less money and success you ever have. Randomness is the enemy. This is not a new success principle. Napoleon Hill included "***ORGANIZED*** Effort" as one of his 17, in *Think and Grow Rich*, in 1937. Today, with business at the speed of light, and so much more to do, use, and manage than in the 1930s or '50s or '80s or last year (!), *disorganized* efforts are not just ineffective, they are deadly. SYSTEMIZING your marketing has moved from luxury to necessity.

Technology can help you build out your marketing system faster and more efficiently than ever in the history of business. I'm reminded of T.J. Rohleder, who has been a very large and successful publisher of business courses and packaged business opportunities for decades. He built his company from scratch to category dominance before we had the tools we have today.

And so Rohleder bought an old hospital to keep everything straight. He would set up different marketing / direct mail pieces in each room and when his employees would come in that day he would tell them which room they would be working in. Today you could have that same kind of precision, and even better, with an average CRM! Today the CRM will just send an email to your printer, or your production team if they are in-house, with a list of people and the marketing piece that they should get. They set it up and run the piece, assemble it, and get it in the mail. No need for a hospital building with 1,000 rooms, giant whiteboards with production schedules on them, physical movement of teams from one project space to another. The computer and all the tools it offers does it all for you!

To back up a step, when we talk about marketing systems and marketing automation what exactly, do we mean by: marketing? It includes:

- Advertising, and coordination of all ad media
- Lead generation
- Converting leads to set appointments or other selling opportunities
- New customer, client, or patient acquisition
- Selling by media replacing manual labor
- Converting first-time buyers to committed customers
- Facilitating upsells, repeat business, retention, and recurring revenue
- Implementing sophisticated, complex multi-step, multi-media communication
- Implementing multi-step follow-up to Call, No Appointment; Appointment, No-Sale; New Promotion, No Response
- Facilitating data mining from within your own records, in order to unearth additional selling and cross-selling opportunities; matching specific offers to specific prospects
- Creating customer enthusiasm and evangelism, and generating more referral activity

- Supporting new product or service launches, seasonal promotions, or other limited-time-period campaigns
- Fully managing ALL online and offline advertising, PR, content distribution, and social media usage
- Creating a lot of "set it and forget it" processes
- And much, much more. Essentially SYSTEMIZING everything leading up to acquisition of a new customer, everything that then happens to keep, fully monetize, and develop a great relationship with that customer.

This is NOT simple or easy. This is NOT a book for people who are lazy or looking for a single magic bullet or a way to get very rich very quick. If you are a serious person, into building a sustainable, successful company and willing to invest both time and money in doing so, you've found the right book! It's not necessary to be tech savvy or even to like tech, incidentally. It's much more important to be or make yourself marketing savvy than use tech or have tech used for you to fully implement your marketing vision, plan, and system. For that, too, you have found the right book.

Now let's get started!

CHAPTER 1

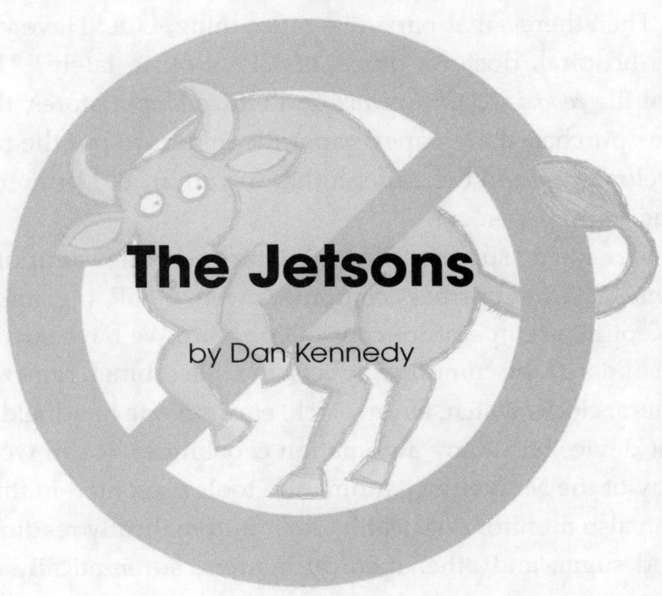

The Jetsons

by Dan Kennedy

Electric, self-driving, flying cars. Robot housekeepers, cooks, valets, secretaries, and personal assistants. The paperless office. Our entire life operated for us by automation. Where is it? We have a future arriving late.

The first electric "aerocar" prototype was built in 1926, and aerocars for all was promised to us in 1940. Individual, personal jetpacks were presented in 1964. Picture-phones were first made to work in experiments in 1956. We saw all this and more in the popular cartoons of the 1960s–1980s. We *do* have sort-of-robots we converse with and get answers from that will also turn lights on and off, operate the TV, order things from the Internet, named Alexa and Siri. The fast-food industry has prototype robot cooks. We may wind up living like the Jetsons, with even more AI-based automation replacing our labor, moving us from place to place faster and easier, erasing many of our human problems. But, from earliest predictions and prototypes, it's slow in coming.

Then there's that paperless office thing. LOL! Have you been in a hospital, doctor's office, or CPA's office lately? They still have file *rooms*. At OfficeMax and Office Depot stores, the main items purchased are paper, paper file folders to put the paper in, envelopes to send the paper hither 'n yon in, and printers to put things onto paper.

We have successfully miniaturized the computer—the phone. My first business computers were 6′ tall, big, and had to be kept in a temperature-controlled room. We have successfully consolidated the computer, typewriter, file cabinet, camera, video camera, clock/watch, alarm clock, etc. into one handheld device. That device has many automation capabilities. It can work with many of the Marketing Automation tools presented in this book. It can also monitor our health vitals, automatically reading heart, blood sugar, and other medical matters, automatically alerting you to your own needs.

We *are* getting there. Late, but getting there.

To Win Business, We Began with Office Automation

When I was a wee lad in short pants, I worked in my dad's commercial art studio, amidst drawing boards, rulers, rubber cement bottles, X-Acto knives, our pet parrot, and the modern miracle of the moment: the IBM electric typewriter. The ads on the pages at the end of this chapter announce the Office Automation that I grew up with, and that were part of my first businesses' offices as an adult.

The Tech Revolution most of you are more familiar with took all the equipment of an office or an advertising agency or commercial art studio and put it inside one box: the computer. Before, in order to have 15 different type fonts, you needed 15 different IBM electric typewriters—a room full. The need for phototypesetting and a dark room ended. For the most part, the drawing board, knives, glue, and paste-up artist, all gone. Today, we've come a long way, baby. Your desktop PC, your iPad, even your handheld device includes an entire building full of Office Automation. Of Production Automation. For good or bad or both, you carry typewriters, paste-up artists, cameras, video cameras, photo and video editing, file cabinets, an encyclopedia, maps, a telegraph machine, endless DIY resources like LegalZoom in place of a lawyer or bookkeeper in place of a payroll clerk, and, gee, I almost forgot, a phone. In your pocket.

Along the way, a whole new kind of automation emerged, too—the one we're tackling in this book, which people in my world call *Marketing* Automation. It began humbly, slowly, unsophisticated. Then it grew like a California wildfire, out of control. Today, it is its own *industry*. Choices are confusing. But if selected and used wisely, it can…

- Erase many advantages big companies long held over small businesses—David can do everything Goliath plus an army can do, what you needed an army for.
- Dramatically boost speed-to-market. The idea you have at 9 a.m. for an ad, marketing campaign, product, or service can be tested for response by noon.
- Virtually eliminate human error in CONSISTENTLY implementing marketing, follow-up marketing, and customer relationship management.
- Facilitate true target-marketing, with fully customized or even personalized marketing communication to different segments of your customer population or overall market—for much greater impact.
- Go so far as to replace a lot of marketing you do (as a verb) with marketing systems you own—liberating you to turn attention to other opportunities and initiatives you otherwise wouldn't have time for, or just "go fishing."

- Make money while you sleep, by converting your marketing operation to 24/7/365, the world we now live in, where potential customers want responses to their expressed interest instantly, any moment, day or night.
- Make your business's sales and customer obtainment and growth more predictable.
- …and more.

For these reasons, even if delving into it and talking about it makes your head hurt as it does mine, you *must* participate.

I Opt Out, but I'm All In

On a personal note, I am what is now called a "modern Luddite." I do *not* own a cell phone or smartphone, my computer is *not* connected to the Internet, there is *no* Alexa or Siri in my home or office, I use *no* apps, I wouldn't do online banking if you paid me, and I judge social media to be a deep, poisonous, mental-illness-creating, soul-killing, dehumanizing, time-wasting sinkhole. I have even more Luddite eccentricities discussed in my book *No B.S. Time Management for Entrepreneurs*. Working on this book required Advil®. If you *dislike* the unleashed Pandora's box of technology, I feel your pain.

But I have it *all* used for me, for my business activities, and for clients. I want this to give those of you flummoxed and viscerally averse to tech and tech talk hope and courage. You can dislike it and still have it put to work for you. You can limit your knowledge and engagement to understanding, but delegate all the implementation. I make a study of it, but I don't *do* it. A chief marketing executive managing online at my client Guthy-Renker said that you need three things to make a powerful, effective sales website: a legal pad, different colored pens, and selling knowledge. If you can draw and explain what you want and how you want it to work, people like my co-author can make it a reality. You need to understand all the opportunities and options available to you with Marketing Automation, which is the purpose of this book. You do not need to personally wrestle with it. And, as you'll discover in this book, there are incredible software options that can do a lot of what you know *could* be done with your marketing if you only had the time, staff, or money. The right-sized software for your business can do it for you, efficiently, cost-efficiently, and effectively. THAT is why all the detail of this book is vitally important to you. It is why I agreed to do it and include it in the No B.S. series of business books.

Rather obviously, I am not the guy to present these details. It'd be like having the best heart surgeon in the world teach you everything you need to know about long-haul truck driving, or vice versa. Thus, the bulk of this book is from Parthiv Shah and several guests invited in by him. Parthiv is a client of mine but also a valued resource for me and for a number of my clients. He will be your capable guide on this journey, sort of Alice and Alex in Wonderland. You'll find that I chime in, in various places throughout.

One last point, an incentive to take this book most seriously. In my 50+ years of working with small- to mid-size business owners, entrepreneurs, and sales professionals who get very, very rich, and those who don't—despite working very hard—I have discovered a single golden thread running through all the successes. It is a golden thread—i.e., theme—of this book, as well as *No B.S. Ruthless Management of People and Profits* and *No B.S. Time Management for Entrepreneurs*. It is: **The less randomness, the more success and wealth.**

Figure 1.1: From Outdated Advertising by Michael Lewis and Stephen Spignesi. ©2017

"Designing a revolutionary concept in software demanded a computer with extraordinary performance. The Tandy 2000 delivered." —Bill Gates
Chairman of the Board,
Microsoft

Bill Gates has been at the leading edge of personal computing from the very beginning. His company is a leading producer of microcomputer software.

"Our newest software product, MS-Windows, is an integrated windowing environment. It will let personal computer users combine individual programs into a powerful, integrated system.

"When we set out to design MS-Windows in color, we knew that the Tandy 2000 computer would let us turn an extraordinary product into a work of art. The graphics are sharp and crisp, and gave us a degree of creativity like nothing before.

"Our engineers were quite impressed with the processing speed of the Tandy 2000's 80186 microprocessor, too. And while the finished product will utilize the 2000's Digi-Mouse, the well-laid out keyboard has helped us speed through the design stage.

"We're proud of our work. So when we want to show someone how great MS-Windows really is, we give them a demonstration. On the Tandy 2000."

Isn't it time you enjoyed peak performance from a personal computer? Go ahead, watch how much faster today's most sophisticated programs run on the high-technology Tandy 2000.

You can choose from the hottest programs around, too, with our exclusive Express Order Software service.

Tandy 2000 systems start at $2999, and can be leased for only $105 per month*. Come in today and see what you've been missing.

Our new 1985 computer catalog is yours for the asking at any Radio Shack Computer Center or participating Radio Shack store or dealer. Check out our complete line of microcomputers—from pocket models to lap-size portables, from powerful desktop computers to multi-user office systems. We have it all. That's why we invite comparison!

Available at over 1200 Radio Shack Computer Centers and at participating Radio Shack stores and dealers.

Engineered for Excellence!
We've introduced the latest in technology for over 60 years. The Tandy 2000 offers twice the speed, graphics resolution and disk storage of other MS-DOS systems.

Radio Shack
COMPUTER CENTERS
A DIVISION OF TANDY CORPORATION

Circle 324 on inquiry card

*Plus applicable use/sales tax. Prices apply at Radio Shack Computer Centers and participating stores and dealers. MS is a trademark of Microsoft Corp

adflip.com

Figure 1.2: 1960s IBM Electric Typewriter Ad

this is the new IBM Electric

IBM

IBM

It's new—inside and out, with 28 engineering achievements that bring you typing at its finest! Your secretary will love its alive, eager response. You'll admire its styling and high-volume output. For this—is the most handsome, efficient typewriter made.

CHAPTER 2

"READY. FIRE. THEN AIM."

(Automate THIS? Automate WHAT?)

by Parthiv Shah

I was drinking some coffee at a local coffee shop, when a man sat down at the table next to me. He saw my laptop bag with the eLaunchers logo on it and asked what it was. I explained that it is the logo of my company and that we help small entrepreneurial businesses all over the world implement and perfect all aspects of digital marketing—including track the marketing so they understand, in real time, moment to moment, which marketing channels are working best for them. Ultimately, we make an automated system for each business.

This guy's eyes lit up and he was full of questions. It turns out he owns a local HVAC contractor and has been trying to make Facebook work for him. So far he said he had failed miserably. "It's been 4 months and I haven't gotten a single call! I'm doing everything that Facebook tells me to do, but so far nothing. I'm really considering just giving up, but I'm not sure what I'll do then because the newspaper ads I survived on for 17 years just aren't producing the results I want either."

I smiled, very sure that I was going to be able to help him fix his issues quickly, and said, "What is your market?"

He responded, "Huh?"

"Who do you sell to?"

"Anyone with a heating and air-conditioning system in their home," he said while looking at me like I was an idiot.

I sighed because right then I knew we were starting from square one. As a matter of fact he hadn't even got to square one yet. He was still getting ready to take the first step.

As a basic foundation, we might start with The Kennedy Marketing Triangle:

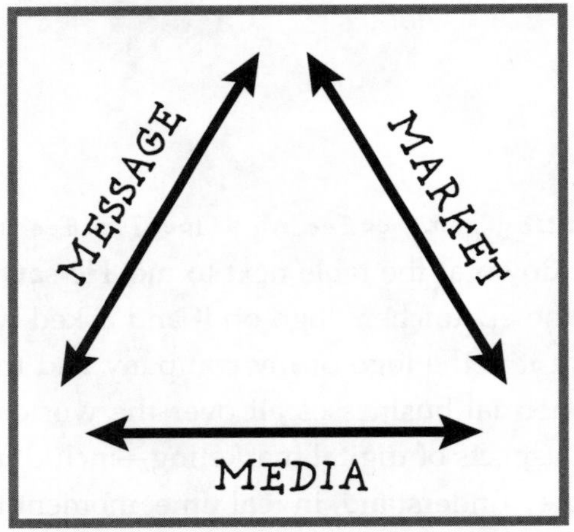

To use it, we have to define exactly who you and your particular business is most appropriate for and can be most appealing to. That may come from your knowledge, from experience, and/or from data mined from your own customer records and/or from other research and/or from a lot of soul-searching. Then we can develop your main messages that match that target market.

Then we can choose the best media and make the best use of it. To give you an example, Dan Kennedy told me about several clients of his in the highly competitive retirement investing and annuities businesses. Everybody in one of their families was a college professor or retired professor. University professors in his state had particular, specific retirement savings plans to be converted to their own investments before or at retirement. He made them his market, he "re-positioned" everything he did to be "for university professors," he used "professor speak," and he used media they read, watched, or used online. All this instead of marketing to any and all retirees, as his local competitors did. By this, he got substantially better ROI on all his marketing. A different financial advisor / annuity company was also generic; for everybody, thus for nobody. In-depth conversations by Dan with its owners, customer data, and other inputs led to determining their best target market was Christian conservatives, married, paying for Medicare Supplement ($300 to $600 a month, not just on Medicare or Medicare Advantage), with modest, mid-level wealth. This, then, governed messaging, media, and made both online/social and offline media more effective.

Now more than ever, you need digital marketing skills to determine: WHO buys from you, WHY they buy from you, WHAT they buy from you, and HOW to approach them. Not just "anyone with a heating and air-conditioning system in their home," but specifically *WHO*. What age are they? What income level are they at? What part of town do they live in? What's their occupation? Much more. Looking for the answers to these questions is not new to marketing, but it has been made MUCH easier thanks to technology. There are tools that address all of these issues, but they come at a cost. As an entrepreneur you need to be a master marketer of your business and have the skills (and time) to interpret the analytics that track website traffic, such as reports, page performance, keywords, and competitors.

Something like Facebook can feed you "perfect" customers or fail entirely, not by its own doing, but by how much you know about your target you are using Facebook to hunt for.

Let's not forget the expert SEO content you are expected to create for your personalized blog posts, and the advanced knowledge of the ever-changing keyword research in your blogs, and blog headings. But that's not all. Email marketing, online products, online sales funnels, landing pages, and social media platforms are a critical part of your branded marketing strategy. You also need to accommodate customers who respond better to direct sales letters and postcards. WHEW! And we haven't even discussed the metrics for offline sales you do.

There is no way that anyone can have all this information stored in their brain. And tracking it with a pen and paper is just too laborious. You would need multiple employees whose only job is to constantly look up these metrics by hand and update them. And that STILL wouldn't be tremendously accurate. Skills aside, how can one person possibly master, let alone remember how, where, and when all the moving parts should be turning to attract, convert, and close leads? Thankfully, you don't have to because a Business Relationship System can do the heavy lifting for you. This kind of structured system includes marketing and sales platforms to help attract visitors, convert leads, and close customers. It can combine a variety of functions that allow marketing and sales activities to be managed all in one place.

The Secret of Repositioning What You Do for Exact Message-to-Market Matching

You have already assembled something to sell. My coffee shop acquaintance had solutions to home heating and cooling, heating and cooling costs, some rooms colder or hotter than others, clean air circulation, and more. But every HVAC has those same things. Facebook, Instagram, other digital media is full of companies offering these same solutions, talking about them generally and generically. To stand out, my acquaintance will need to reposition his solutions and find ways to talk about them differently, even to the point of being proprietary. Help with this can come from the WHO he decides on as his best, most desired customer. Taking a "generic" out into the marketplace, using every available digital marketing platform for it, being one of many in the same places, is a losing proposition. You have gotten ahead of yourself. Spending money and time BEFORE having a defined target and BEFORE having offers specifically appealing to that target.

Just as quick examples relevant to my coffee shop acquaintance's HVAC business:

1. Save $200 to $300 a Month on Your Home's Heating/Cooling Bill!
2. Fund Our Son's or Daughter's College Fund with the Savings I'll Get for You, on Your Monthly Heating/Cooling Bill!
3. Enjoy Your Retirement More—Pay for Two Great Vacations or Cruises a Year with the Savings I'll Get for You, on Your Monthly Heating/Cooling Bill!

#1 is generic. #2 and #3 are targeted. If we can match #3 with relatively young retirees of modest means, who own their own homes...if we can match #2 with parents of high school–age kids, who own their own homes, and are mid-level income earners... or if we can use these separate appeals, "fleshed out," in social media to attract the matches...we can be much more efficient and effective. Now we have something worth applying the full force of Marketing Automation to. In fact, we have *two* such things.

Case History: Get More Patients, Better Patients, at Lower Costs

Let me give you an example with one of my clients. They have a multi-seven-figure clinical practice with a great reputation in the market for clinical excellence. Patients travel from long distances and check in to a local hotel just to see the fine doctors at this clinic. The clinician is a highly educated individual with significant investment in continuing education. He is an avid reader, astute student, and a prolific writer. He has a great "chairside manner." His staff is excellent. Despite all this, he was in a constant struggle to attract enough suitable patients, and found it nearly as difficult now as it had been years ago. His question was: Shouldn't this be easier and less stressful by now?

The difficulties with sufficient new patient attraction were not inside the office or with him. They were with the media representing the practice. For starters, their website (online practice) ambience did not match their physical clinic. It was incongruent. The website infrastructure was poor. Capture of visitors' contact information and data was ineffective—many disappeared back into the water like slippery fish. Automated follow-up to site visitors who did not immediately book appointments was nearly nonexistent. Data was not used to fashion messaging. Quite a bit of money was being spent on media like Instagram, Facebook, YouTube, etc. to bring people to this site, but it just wasn't ready to perform.

We did a deep dive of what the goal of the Pixel Estate (website and online media) was as well as who the ideal patient is. We mined his existing patient data for clues to the ideal patient's demographics and psychographics. Frankly, we knew more about his patients, ideal patient profile, and many facts about his practice within days than he knew from years. Using this information, we built a much more targeted, appealing, and functional site using

HubSpot (one of a number of technologies you'll be introduced to in this book), allowing for automation, direct downloads of a consumer's guide to choose a pain management doctor, easy access to other information, video presentations, and direct booking of appointments. We also better connected all the digital media he was using.

With all this, we achieved much better prospective patient engagement, ROI on dollars spent to get them to the website, and the boost in overall percentage of people who became patients was dramatic!

We, incidentally, have a version of all this work combined with ready-to-use marketing just for implant dentists. You can see it at SellMoreImplants.com.

When you have a consistent, trackable, well-thought-out process designed to engage, build trust, and excite prospects about working with you, you end up with lower acquisition costs and increase profits almost automatically. Most business owners have not accomplished this. They have hurried into "digital" with Ready, Fire, then Aim. They have disjointed, disconnected presences in different online places and platforms. Frankly, their digital marketing is a mess! Their offline marketing is usually also disorganized. Hopefully, here, I've motivated you to want to fix this. In the next chapter, we'll start…

CHAPTER 3

David vs. Goliaths, Punching above Your Weight Class and Going from Confusion and Chaos to Control

A Patented Growth Strategy System for Small Businesses in the Digital Age

by Parthiv Shah

Small business owners today face an increasingly complex and competitive landscape. With the rise of digital technologies and online platforms, customers have more choices than ever when deciding where to start a buying journey, where to get information, and where to spend their money. At the same time, larger corporations with big marketing budgets put constant pressure on small businesses fighting for market share. Most small business owners struggle to keep up. They know they need to have an online presence and use digital marketing channels to reach new customers. But the vast array of technology options, from websites to social media to email marketing, and more, can feel overwhelming. On top of that, few small business owners have extensive marketing knowledge or large budgets to spend on a big agency.

The key is finding the right growth strategies that align with your core business strengths while leveraging smart digital tactics. With the right strategic approach, you can use today's marketing technologies to punch above your weight class, despite limited resources.

The Growth Strategy System is a unique, patented system and method of studying historic revenue and other internal data to identify reasons why a brand or business has been growing, who its best customers are, why those customers were attracted to it, and use that data as a model to create a customized business growth strategy, targeted messaging and media, and ultimately automate it all. In November 2022, U.S. Patent #11,507,967 was issued to Parthiv Shah for this invention. That's me! I don't know if you hold any patents, but securing one for a marketing process is very unusual and rare. This speaks to how unique my methods are.

The system will increasingly use the power of "Predictive AI" to analyze the current market data while using the power of "Generative AI" to generate each customized growth strategy. The Growth Strategy System is being developed as a SaaS software, and someday will become a tool for agency owners, consultants, and brand CEOs. I've used the system with hundreds of clients and thousands of projects.

The system guides owners to evaluate their business across six key areas and identify the best marketing tactics to pursue. Rather than throwing money at the latest marketing fad or newest platform, the system takes a more calculated and strategic approach.

The six elements of the Growth Strategy System are:

1. Brand Awareness
2. Lead Generation
3. Sales Conversion
4. Client Retention

5. Client Referrals
6. Revenue Growth

BRAND AWARENESS MADE SIMPLE is how immediately recognizable you and/or your company and its core promises and main story are to and within your target market or audience. If 100 people in your market are picked at random, given a generic description of what you do or the solutions you provide, and asked to provide a specific business by name—how many would name you as the first to come to mind? Would you score 80 percent? Fifty percent? Uh-oh, 10 percent? How many people in your market will bring you up by name in any occurring conversation about the solutions you provide?

LEAD GENERATION MADE SIMPLE is the quantity and quality of new, interested potential customers coming to you as a result of your advertising, marketing, PR, word-of-mouth, and other sources, at what pace, and at what consistency. Do you have a system or systems working for you that generates a satisfactory quantity and quality of leads, for follow-up, to convert to customers? Online, incidentally, this is not measured in visits or views or likes, etc.—it is measured by the number coming into *your* funnel and by the appropriateness of those entering your funnel, appropriateness meaning they match up well with your known criteria for a desirable customer.

SALES CONVERSION MADE SIMPLE is the percentage of leads that engage in an initial transaction. That buy. This, incidentally, must be tracked by source. As an example, in examining one marketer's data, he had 50 percent of his webinar attendees coming from one source, only 30 percent from the next best, then single digits from others. This was as deep as he looked to score the first source "the best." But only about 20 percent of the 50 percent bought, while almost 60 percent of the 30 percent bought. You might play with that math.

CLIENT RETENTION MADE SIMPLE is the number of months (or years) a customer or client stays "active" with you, however you define "active." This, too, has to be tracked by source and by other variables, such as different initial transaction, initial price, and ascension.

CLIENT REFERRALS. Most business owners take what they get. Little is done to organize and manage a system for maximum possible referrals. For this, you should get the *No B.S. Guide to Maximum Referrals.* Dan Kennedy taught me one way to measure your success with this, which he got from Paul J. Meyer: the percentage of customers you create with what Paul called "endless chains of referrals." With an endless chain, each customer provides at least one referred customer, and that customer provides one, etc, endlessly.

To be a necessary broken record, a lot of what you can do to improve measurable results in each of these six categories can be automated, via your CRM (Chapter 6), your marketing software (ClickFunnels, Keap, HubSpot), and other tools. To do everything that can be done to improve results in these six categories, it will have to be automated.

The power of the Growth Strategy System comes from accurately assessing your business across each of these elements, developing integrated strategies to capitalize on strengths and overcome challenges, then closely tracking execution. Everything you do or use or spend on becomes measured and data driven.

My friends at Keap label this "conquer the chaos." I find that, by working within each of these six categories of marketing, getting relevant data, setting up scoreboards, and measuring results in as close to live time as possible, we can replace guesswork and stress with knowledge, accurate forecasting, and organization.

Visit www.meetparthiv.com to learn more about the Growth Strategy System for small businesses.

Figure 3.1: Patent for Growth Strategy System

Growth
STRATEGY
System

PATENTED

US Patent No.: 11,507,967
November 22, 2022

System and Method for Developing A Growth Strategy

Parthiv Shah
President
eLaunchers

launchers

www.GrowthStrategySystem.com

Dan Kennedy Says

Every pro athlete, coach, and business consultant being honest will tell you that measurement *automatically* improves performance. Often, if you change nothing going on in and with a business, but measurement of every "little" thing, you create documentable improvements.

There's a famous story involving Andrew Carnegie's steel mills and a consultant, Charles Schwab, asked by Carnegie to somehow boost production of each of the three shifts working in the plants. He did so with a fantastic "technology": chalk. At the end of each shift, the shift manager wrote the number of their production on the floor, big, in chalk. If the next shift beat it, that manager could erase the old number and write theirs. If they fell short, they had to leave the prior shift's number there and write their own. Production became highly competitive. Most sales managers, where a number of inside sales reps work, by phone or seeing customers in person, keep a scoreboard up on the wall, and move each rep up or down, above or below each other, every day.

You can establish quotas, benchmarks, and other "scores" for multiple items in each of the six categories of achievement in a business that Parthiv likes to focus on, as well as others. You can have the scores kept by automation instead of by manual labor. The more you do this, the more successful your business will be. You should be getting a scoreboard spit out by automation put into your hands every day. Just as an example, you're no longer limited to evaluating tele-sales reps just by their conversion percentage. You can score their sales efficiency, too, with automated readouts of how many minutes they spent on each call, the number of minutes in a day actually on calls, and other facts. Each person they book to an appointment or that they sell to can be "tagged," automatically followed, so you see differences in the ways that John Rep's buyers behave over 90 days vs. Jim Rep's. Jim might have a slightly lower conversion rate, but keep more of his appointments, or more of his appointments become more engaged with sequential purchases than John's—so Jim is actually the better of the two. Source also often matters a lot, so John and Jim may have better conversion rates with people attracted through Ad Media #A vs. Ad Media # B, in which case you'll want to consider ways to increase spend with #A, decrease it with # B, and go hunting for another media that works like #A. This dynamic exists with many things *not* being scored and evaluated.

I have a client that sells via webinars, to members of a certain industry. There are 11 usable ad media, and for follow-up, four in-house tele-reps. By carefully evaluating and using a lot of automatically provided data—for example, what percentage of those from each of the 11 media stay on the webinar until its end vs. exiting early—they have been able to boost sales conversions of the webinar by 16 percent and boost productivity of the reps by 7 percent over the course of a year. This equates to well over $300,000.

A good exercise is to think about and list every item of information you could use if you got it, that you don't now get. Then set about getting it!

CHAPTER 4

We Begin to Fish Using Automation and AI for WARM and COLD Email Outreach

by Parthiv Shah

Maybe the most basic tool of Marketing Automation is email, so I decided to talk about it first, ahead of other important aspects. You know you can use it to stay in touch with your clients, patients, prospects, but can you use it as a very inexpensive and effective outreach tool to lost, past customers, and even to first-time contacts, strangers? Most of my clients are used to HubSpot and Keap (formerly Infusionsoft). With these tools, there are rules. We cannot simply buy a database and upload it to HubSpot—I'll get kicked out like that. It's a problem. Also, if we get mass email sent, deliverability is a hurdle, getting classified as spam is a hurdle, and the open rate is a hurdle. Still, there *are* ways to effectively use email.

One of the things my friend Seth Greene discovered was if he sends a single email, from his Microsoft Outlook account, you will get it. It will land in your inbox because it's a personal email coming from his email account to your email account. It isn't sent by HubSpot or Mailchimp. He figured out that if he could send one email at a time to one person, he could get 100 percent deliverability, he could get an exponentially higher open rate, and geometrically increase his results.

For me and my business, Seth figured out that I could send one-on-one invitations to a webinar and I could get a lot more people to show up than by any kind of mass marketing. Even the reminders that come from Zoom, the day before and an hour before, are ignored by a lot of people, but a seemingly personal email is different. I also sent follow-up emails to webinar attendees, saying, "You didn't take advantage of the offer we made on the webinar," then making another attempt at that sale. My email didn't land in promotions, it didn't land in spam, you got it in your inbox.

It is my friend's 16-year-old son Max's computer code doing the work. He isn't physically sending tens of thousands of emails. His bot is mimicking the physical actions you would take to do that.

Lost 'n Found Money

You have unengaged people on your list. Unengaged means they haven't opened an email in 30 days, 60 days, 90 days. HubSpot, Keap, Mailchimp, they all automatically stop sending your emails to people who are unengaged, because doing that would affect their IP sender reputation. If you just let this be, you have an ever-growing number of people never receiving communication from you again, even though today's disengaged past customer or unconverted lead can be tomorrow's interested prospect. I exported out my Keap list of unengaged people. I used the bot to send them an email, they actually got the email, I reactivated a lot of those leads, and actually brought in business. They said, "I wondered where you were; I hadn't gotten an email in a while." I said, "That's because you didn't open them. Keap automatically stopped sending them to you."

My friend Seth Greene was literally spending hours a day working nights at home, upsetting his wife by always working, sending one email at a time. He would go copy and paste hundreds and hundreds of times, an extra couple hours a night before a webinar, to get this done via manual labor. He even went downstairs after his wife fell asleep. It was a lot of manual labor. His 16-year-old son asked a life-changing question: "What can I automate in your business that is working when you do it manually?"

That's a million-dollar question. Dan calls it "a writer down-er." He'd say: Write it in big, bold letters on a 4" x 6" card and tack it up on the wall where you'll see it a lot. If you have staff, give a copy to each staff member. Tell them to tell you what they do manually that drives sales or retention or referrals that might be automated, liberating them for another opportunity.

Seth said, "If I send one email at a time from Outlook, I can get through and it works. Can you automate the physical manual labor of sending one email at a time?" And his 16-year-old figured it out. Invented a bot that could fool everybody, and seem like an individual sending one personal email to another individual. We have used this to send 600,000 emails. When you send 600,000 emails, you learn a lot about what works and what doesn't work. I am going to give you some priceless lessons in email marketing that, whether you use the bot or not, you can apply in your marketing right now. Then I'll let you use my bot, if you like. Here are the lessons we've learned:

- When it comes to subject lines, don't use all caps—it decreases your open rate. The server's filters delete these.
- Keep it short. If you keep it under 50 characters, you have a better shot of getting opened. If it is going to be long,

you must know it is going to a highly interested audience responsive to you.

- Save the emails you get. We have one of the best swipe files of proven results, getting emails. There is a priceless training inside your email inbox.
- Robert Collier said you need to participate in the conversation that is already going on in your prospect's head. If I were promoting a dinner with Dan Kennedy to people who are going to Funnel Hacking LIVE, then I would talk about Funnel Hacking LIVE—I would talk about things that were going on.
- Ask a provocative question. One of the ways to increase your response rate is ask them a question and get them to reply to you. Anytime someone replies to your email, that increases your email sender score because it looks like a conversation.
- Use specific results and the time it took you to get there. This email bot can deliver 360 emails an hour. It gives you results—360 emails in 60 minutes, delivering emails every 10 seconds. Stop. Evaluate. Adjust. Split-test.

I am even going to tell you our best performing subject line in about 63 different industries. It works business to business. It works business to consumer; it does not matter. We have tested this subject line in every campaign we run. It's one of our split-tests. People are social creatures—they want to feel like they're not alone in the problem they've got. They want to feel part of something. I'd be really surprised if it didn't work better than whatever your normal subject line would be. Here it is: "You Are Not Alone."

Our second-best one came from Dean Jackson. He has a very famous nine-word email—"Are you still looking for [blank]?"—and the blank is whatever you do. Are you still looking to get top dollar for your home when you sell? Are you still looking for a way to really lose weight fast? Are you still looking for a way to find better employees? Are you still looking for blank? It's an immediate lead qualifier. Will everybody reply to your email? No. Every time we send this, though, we get people who reply and say, "Yes, I am still looking for," and then you've got a qualified lead.

You've Heard of (Illegal) Money Laundering. Here's (Legal) Lead Laundering!

Now let's talk about some of the use cases. We've talked about how you could use this to reactivate an unengaged list—individuals who haven't replied to your email in one month, three months, six months, nine months. You can deduce they're probably not even seeing your emails anymore. Use this to get them to be reactivated.

You can even use this to create lead flow from cold sources. We've talked about buying a list of dentists and how you can't upload it to HubSpot. However, you can use this bot to email them all directly, and that could drive them to your HubSpot landing page where they've now given you permission to communicate with them. This is usually done with a Lead Generation Magnet offer, like a free report download, free book, free training. They are now a *clean* lead in HubSpot and they can go through the HubSpot automation sequence and you can now drip on them until the end of time in HubSpot. If you want to fill your HubSpot with more qualified, interested people, use this bot to get them to go opt in for something through HubSpot, etc. (Or other software like Keap, or moved into ClickFunnels.)

For me, this significantly increases webinar attendance rates. We did the exact same webinar twice and we had it with the same webinar partner. The first webinar we did without the email bot, just the Zoom automations reminding people to show up. We had 814 registered and 244 people showed up: 30 percent. We had 814 people registered the second time in. We used the email bot to remind them to show up and 454 people showed up: 55 percent We've used this to book podcast guests. I scraped an email list of 1,000 estate planning attorneys off of LinkedIn. I then used this bot to, in 2 hours and 45 minutes, send 1,000 emails to attorneys we didn't know, saying: How would you like to be on a podcast about estate planning? We ended up with 125 podcast guests in two and a half hours with no manual labor. Our podcast producer literally said, "Wow, you got me 125 guests. We were airing once a week—this was going to take 2 years to get through. We're now airing every day." She's said, "Please don't do that again until I catch up." That's a wonderful problem to have.

You may or may not use webinars or podcasts in your marketing. Maybe you should. Maybe you shouldn't. But I hope you can see that my experience with this for my business and for clients can adapt to do something significant in lead generation for your business.

The implementation of the email bot, while innovative and effective, was not without its challenges. A significant aspect of its development and deployment involved navigating the complex landscape of email compliance and addressing various technical hurdles. One of the foremost challenges was ensuring that the email bot complied with various email sending policies and spam regulations, notably the CAN-SPAM Act. This act sets the rules for commercial email, establishes requirements for commercial messages, and gives recipients the right to stop receiving emails. To adhere to these regulations, the email bot was designed with built-in features that automatically included an opt-out option in every email and managed unsubscribe requests efficiently. This ensured that the emails sent were not only personalized, but also compliant with legal standards. Another technical challenge was managing bounce-backs and opt-out requests. Bounce-backs occur when emails are sent to invalid addresses, and managing these effectively is crucial for maintaining a clean email list and a good sender reputation. Max coded a second bot with the capability to automatically update the mailing list by removing invalid addresses and processing opt-out requests promptly. This feature helped maintain the integrity of the email campaigns and ensured respectful communication with recipients.

If you want to take the Invisible Email Automation Bot for a 14-day free trial, simply go here: www.elaunchers.com/email

Dan Kennedy Says

In direct mail, there are long-lived strategies for "sneaking up." One of the G.O.A.T. copywriters, Gary Halbert, observed that "a lot of people sort their mail over the wastebasket. You have to survive the cut." If you look like junk mail, you risk being handled by postal workers as junk that can go undelivered without consequences. There's a junk mail / bulk mail non-delivery problem just as there is an email deliverability problem. You risk going directly to the recipient's wastebasket. So, a lot of direct mail is designed to mimic personal, one-to-one mail. As an example, a marketer I spent some time with was very successful with a membership offer, sending his sales letter folded, put in a greeting card–sized envelope, hand-addressed, sent First Class with stamps, and with a mark on the sealed flap of a woman's kiss in red lipstick. Yes, he had a huge room of women seated at long tables, stuffing, sealing, hand-addressing, and planting kisses on the envelopes. Each looked like it HAD TO HAVE BEEN SENT by a person you knew, personally, to you. I copied this approach (without the lipstick mark) very successfully for a skin-care product, for WeightWatchers International, and for a B2B campaign to CEOs raising funds for restoration of Ellis Island, from Lee Iacocca. What Parthiv has described here for email is this kind of direct mail. It is the same trick.

All advertising and marketing currently suffers mightily from the massive, monstrous amount of clutter online, existing because the media is cheap or even free. This makes your mandate doing something different, differently, so you can avoid the frantic dumping of 99 percent of what is seen or arrives digitally into the mental or virtual wastebasket.

CHAPTER 5

When They Arrive at Your Digital Doorway, Then What?

by Pathiv Shah

Imagine meeting Steve Jobs when Apple was in its infancy.

I have to start this by telling you a story. I showed up bright and early for Dan Kennedy's big, multi-day Super Conference. In 2015. I was a premier event sponsor, and this is what happened.

I had been working the room all day and into the night. My feet were tired. My body was still feeling the effects of jet lag. But my mind was so full of ideas from the day's event that I knew sleep would not come anytime soon. It was around 9:30 at night when one of Dan's team members, Nick, tapped me on the shoulder.

"Hey, Parthiv, tomorrow's keynote speaker just arrived. His plane was late. He's sitting alone in the bar. Why don't you go say hi and give him some company."

So, instead of heading to my room, I followed Nick into the bar. The place is relatively empty, but as promised, I saw a young man sitting at the bar alone, sipping a glass of water. I stop in my tracks. Surely, this all-American-looking young man can't be tomorrow's keynote speaker. He looked like he belonged on a high school baseball team. Nick leads me straight to him. "Russell, there's someone here I'd like you to meet." He proceeded to introduce me as a premier sponsor of the event. The three of us shared a bit of small talk, and then Nick says his goodbyes and left us for the night.

Now imagine… There I am, it's 9:30 at night, and I've just met *Russell Brunson*. At that time, I had no idea what a big deal this was or the impact it would have on my career in the coming years. "So, what do you do?" I asked, hoping for a preview of what tomorrow's presentation might be about. He smiled. He grabbed a pen out of his pocket. He reached for a napkin out of the plastic holder on the bar. And he started drawing stick figures. If I had only known what I was witnessing then, I would have kept that napkin. Russell's stick figure drawing would be worth a fortune today.

He started to tell me about a thing he had named "click funnels." He was illustrating with his stick figures while he spoke. He told me that the funnels he had invented were like digital roads that move your dream customers to your website with credit cards in hand. Once someone enters, they are "funneled" along whatever sales journey you design, contained, unable to roam free—like water poured into the top of a metal funnel.

He talked about how it was a one-stop marketing shop. A shop that gives you everything you need to market, sell, and deliver online products or promote any kind of offline business online. He said his slogan was: You Are Only One Funnel Away from Your Fortune.

I remember my excitement building as he talked. In part because he was so passionate about what he was saying. And in part because I was already realizing how I could use his funnels to serve my clients.

We talked on. Ten minutes' and two napkins' worth of drawing later, he asked, "What about you, Parthiv? What do you do?" I told him I design and build landing pages and personalized URLs. He shook his head and chuckled. "Aren't you afraid my software will make you obsolete in a few years?" I grinned. "Nope, not in the least. And here's why. From our 10 minutes together here and your stick figures, I can see this is powerful. A game changer. Tomorrow, you're going to walk onstage. You'll tell the audience all about the power of ClickFunnels and how they'll make a lot of money once they buy your software. They will buy it." Russell looked at me, puzzled. "I don't see how this works in your favor, then." I smiled. "My prediction is everyone who buys your software tomorrow night will hire me in the next three to six months. Most of the people in the audience are busy. They are busy being dentists. They are busy being doctors. Lawyers. Real estate developers. Many are thrilled that they can turn on a computer and access their email without tech support. They will see the potential of your software. Recognize it as the innovative next step they must take to be successful. Then promptly feel overwhelmed when they have to read the owner's manual—even though you make it simple. They are simply too busy. And that's where I come in."

Russell smiled, giving me a knowing look. "I see your point."

I pointed to his stick figures. "I get what you're talking about. I think I'll be one of your biggest fans. Tomorrow, I may be the only guy in the room who *really* understands what ClickFunnels can do. Your software will help me go faster. It will help me serve my clients better. It will help me make more money. And that's why I'm not worried. Because ClickFunnels will help me achieve my goal—helping my clients build an online presence that makes them a lot of money."

My business has become much more complex and multifaceted since then, just as the digital world has. The secret to success then and now and for the foreseeable future is being able to fully organize and control the thoughts, interests, movement, and maturation of a potential customer, from the very first sight of you they have, wherever that is, forward all the way to the sale. You might want to pull that sentence out and write it down somewhere. Next, the secret is to be able to actually implement the scheme you architect to do that. That is what my business does for businesses now, and ClickFunnels is a big tool in our tool kit.

The next day, I attended Russell's session. He was on fire. The audience loved him. His passion and engaging manner were infectious. I was among the first in a long line to buy ClickFunnels. It was one of the best decisions I've ever made. And my life has been changed ever since.

Russell, if you aren't familiar, has become a major business celebrity in the online marketing space. Maybe, arguably, the Steve Jobs of the field. His company has tens of thousands of users, an army of consultants using ClickFunnels for their clients, and a complete ecosystem for support of his users. His annual ClickFunnels Live is attended by over 5,000 people.

Today, I have over 450 Click Funnels. I build more all the time. Five of my funnels have generated well over 9 million dollars. directly, and much more over the long term of the customers' lives. As I write this, I've been a ClickFunnels Certified Partner for almost 9 years. I've won Two Comma Club awards—the award granted by Russell Brunson to those who have generated over a million dollars in revenue with a single Click Funnel. I've done this twice. And I have also won the prestigious 2CCX award, granted solely to entrepreneurs with a single Click Funnel that generates over $10 million in revenue. You can see my journey with ClickFunnels from 2016 through 2019, including how I earned these awards at eLaunchers.com/journey.

Is It Really True—Can You Be Just One Great Funnel Away from Your Fortune?

What took me over the top…what got me where I am today with ClickFunnels—award winner, record breaker, moneymaker—was when I took a book funnel from Russell's book *DOTCOM Secrets*. (By "book funnel," I mean a Click Funnel that promotes a book, which, in turn, promotes a business. Something any marketer of anything, from a local dental or financial advisory practice to a global marketer of nutritional supplements, can use. It is *completely* automated.) I studied it. Went over every nuance. Implemented it. And started driving traffic to it in 2016. I've been growing rich off it ever since. In 2019, this funnel made its first one-millionth dollar. This won me my first Two Comma Club award with ClickFunnels. I was excited. And I wanted more. I could see the potential. I could see that I'd barely touched the tip of this iceberg. I knew that if anyone could help me capitalize on this idea, it would be the man with whom I'm writing this book. I went to Dan Kennedy. He augmented my book funnel. With his guidance and advice, I made it better. I made more money. The cash rolled in. I attached my "Business Kamasutra" marketing automation campaign to the back of the funnels and created an end-to-end implementation of Magnetic Marketing. Things were going great…better than I'd predicted. *Then the world fell apart.*

I think this story is important because there are occasional black swan events. It's part of business and part of life. You can best survive them financially if you have good, productive systems in place.

Do you remember what was happening in April of 2019? Neither do I. But I bet you do remember April 2020. Glued to the television. Listening to the talking heads in Washington and our local government. Stay inside. Wear masks. Avoid everyone. You likely remember the rush at the supermarkets. The rationing of essential items like toilet paper and cough medicine. How every time someone coughed, it started a panic. Did you lose a loved one? Did you lose your ability to earn a living? Did you lose your business? 2020 was the year business as usual *stopped.*

You are not alone. In 2021, Vice President Kamala Harris noted that over ⅓ of small businesses closed permanently due to the pandemic. As I write these words, the numbers are improving, but the hit was brutal nonetheless. COVID sent me scrambling. After all, I am a small business owner. During one of Russell Brunson's seminars, I remember he posed the question, "What will you do if everything is taken away from you?" COVID tested this. My revenue dried up like the African Nile during a drought. What was I to do? I had a team to feed. A family to feed. A business my wife and I had put our hearts and souls into.

I remember Russell's question swimming through my head every day. "What do I do now that everything has been taken away?"

I made a set of promises to myself and my team. No one gets fired. No one has their hours cut. No one's hourly rate is reduced. Vendors get paid on time.

Spoiler alert: I made it through. How? I called in the big guns. I called Dan Kennedy. I called my chief copywriter, Russell

Martino. I made a wish. "Make it rain, guys! The sky has fallen. It's time for the battle of our lives. It's the day we've been preparing for."

They delivered. And out of the darkness came the light. My new Prism Campaign, with its subsequent funnels, was born through the collaboration and genius of Russell Martino, Dan Kennedy, and myself.

The Prism Campaign saved eLaunchers from going the way of so many unfortunate businesses during the height of the pandemic. My promises were kept. Not only was I able to pay and retain my staff and pay vendors on time, but by August 2020, Prism had generated enough revenue to get us out of the red, and by the close of 2020, we ended in the black. Despite the struggle, heartache, and anxiety, we ended 2020 as a profitable year. Without having systems and automated systems, we'd have died.

But the story doesn't end there. The Prism Campaign made its one-millionth dollar in early 2021, earning me my second Two Comma Club award with ClickFunnels. This campaign pulled us through the most challenging year in my professional career. It then surpassed my expectations and set a new precedent for how we do business at eLaunchers.

The Prism Campaign was all about "sharing—not selling." I told business owners how my funnels worked and then gave them away for free. This worked for the same reasons I discussed with Russell Brunson at our first meeting long ago at a lonely bar counter far, far away. The funnels work. Anyone with a brain can see that. Giving them away for free shows I am willing to offer something of value. In a time of trauma for most business owners, I chose *not* to ask for up-front fees or sell goods or services at the start of new relationships.

Still, my clients are primarily doctors, dentists and lawyers, real estate developers, and some peers in what we call the information marketing industry. These businesspeople are busy. They can tell my funnels will make them money. But they don't have the time to implement them. So, they called me. Mr. Implementation. That started many conversations that led to profitable, long-lasting relationships. Relationships from which my clients and I still profit today. So here's the moral of the story. ClickFunnels works.

CHAPTER 6

Your CRM System, Your Essential Automation to Sell More to More Satisfied Customers Who Will Stay, Pay, and Refer

By Parthiv Shah

Seducing a single person is a lot of work, but it is doable. However, enticing hundreds or thousands of people to choose you, sign on with you as a repeat or recurring customer, set aside lowest price interest, and stay in love can be overwhelming! Trying to do *everything* needed for such strong relationships by yourself or manually by your staff is impossible. From time to time, there won't be enough hours in the day to send those thank-you notes or recognize those birthdays or get out a commentary on news important to your customers or promptly acknowledge a question or problem. Decided on relationship enhancers, things, people will slip through the cracks.

This is where technology comes in. Specifically, a customer ***relationship*** manager (CRM). There are many CRMs available to business owners, but they all have one thing in common—their entire purpose is to help you establish a long, fruitful, strong bond with your customers, clients, or patients. It is not just to manage lists and data or provide a necessary evil of customer service. It is to build relationships.

Over the next few minutes, I am going to share some generalized information about how you can use CRMs to massively increase your interaction, sales, and customer value. Before we get started in earnest, let me share the role of data in your client, customer, and patient relationships.

Why and How to Get Data for Your Relationships

Data in relationships is information and feedback. What you *don't* know about people you have relationships with can get you into a lot of trouble. What you do know can be extremely useful. In an individual relationship, things like her attitude about Valentine's Day, favorite color, meaningful childhood experiences, food allergies, love of (or dislike of) old movies, etc., are all useful. You collect this without thinking about it as *data*. But it is. With large groups of customers, this same kind of data is hard to come by and more difficult to use, although list segmentation by our CRM and/or with your Keap, HubSpot, Mailchimp, or other software is doable. But with big customer populations, you will probably want other useful data. Do these customers only respond to "big sale" promotions, while these other customers aren't nearly as responsive to such promotions as they are to V.I.P., By Invitation Only Exclusive Trunk Sales? Do these customers have kids, but these customers do not? And so on. This data can facilitate both list segmentation and on-target messaging. So, we want to:

1. Capture, catalog, analyze, and display data that builds a relationship.
2. Use data to manage friction in relationships. If you know they think about "x" in a certain way or have a particular

set of references related to "x" (by their age, occupations, incomes), you can avoid innocently irritating, and you can leverage agreement. For example, if you know and can segment which of your customers served in the military and which didn't, you have a good opportunity for a Veterans Day promotion delivered only to veterans.

3. When you are able to talk to different groups within your customers differently, at different times, you can be more effective.
4. Whenever you default to mass, generic communication, you reduce the feeling of a relationship. The more you communicate to Mary Customer in a way that shows you know her, about her, and care about her concerns, the more you have a relationship.

The good news is that we can use a CRM to automate those systems. That way we can easily build trust with our prospects—as long as we are attracting the right prospects to begin with.

The First Date Is Just the Beginning

A lot of businesses foolishly, lazily, or ineptly treat their first transaction as the victory at the end of a certain amount of work and investment. Oh boy, we made five sales today! Really smart, industrious, and capable business owners think of that same sale as the start of something.

Let's assume you had your first transaction with your client. Congratulations. I hope it was as enjoyable for the customer as it was for you! I hope it was profitable. I hope it is the beginning of a brand-new relationship. The first transaction is important because it can put you in a role, with status, where the customer is getting accustomed to trusting you and willing to pay attention to you whenever you show up, however you show up. That will be cemented or it will dissolve depending on everything you do next—not what you did with the first sale.

If you scheme out everything you can to convert that first date into a situation where the customer is just as eager to see more of you as you are eager to see more of them, you then have the challenge of no-fail, consistent implementation. Even if you have a system in your head, if that system isn't being executed with the same dogged reliability as a robot tightening a bolt on an assembly line—exactly to X torque on every piece, every time—*you might as well not have a system at all.* Business owners have a lot of good ideas and plans and schemes in their head that don't happen. That's where a CRM comes in. With a CRM you can set up automation that will make sure every customer is handled the right way every time.

If they buy X, they get one follow-up letter and email sequence. If they buy Y, they get a different follow-up letter sequence and only one email acknowledging the purchase. If they buy one product or service, we know they are good prospects for something else...so we automatically market to them. The CRM will tell the print and mailing house what to send to whom, when. It will send emails to your clients and then follow up based on their actions with the email. Did they open the email? Did they click on a link? Did they reply? It will tell you. If you need someone to follow up with a phone call, the CRM will notify the correct person in your office and continue to harass them about it until the call is made. If you collect data, the CRM will sort their names and contact information into different groups you set up. Whatever needs to be done next to strengthen, deepen, and reinforce the bond between you and your new customer/client/patient can be programmed into the CRM. And the CRM will see that it gets done.

If you currently have a CRM that is not doing all these things for you, you may want to investigate all its capabilities and improve your use of it or replace it and/or get outside, expert help with this. Different kinds and sizes of businesses need and can profit from different levels of complexity and sophistication in their CRM. For some very small businesses, Constant Contact is sufficient, but for others it is utterly inadequate. Many can use Keap and/or ClickFunnels. Some need multiple systems interconnected with HubSpot. You should take the time and trouble to diagnose all your needs and opportunities, evaluate different softwares, and get the best toolbox for your purposes. Often, having "fresh eyes" do such diagnosis for you, and having software selected and put into operation for you, is a good option. With it, you buy time and speed to success. If that interests you, feel free to email me at pshah@eLaunchers.com.

Movin' On Up

We want to keep customers, make them more active, and possibly ascend to higher price tiers or higher levels of connection, membership, or activity. The design and use of a good Ascension Ladder is a subject too complex to include here. You can see a good example of one at MagneticMarketing.com. It's important to know that ascension is the best form of retention.

In my business, at eLaunchers.com, we want all these things: more engagement, use of more services, retention for a long time, and ascension to higher levels of relationship. Keep in mind, this entire process is programmed into my CRMs (we use multiple ones to manage all of our data). That means everything that I am about to describe is done automatically.

You start with a free report—like the Success Blueprint on my website. As soon as you request it, I am going to take you to a thank-you page. You are going to receive your digital copy of the Blueprint, and then I am going to say, "I have a couple dozen mind maps of my notes along with some workbooks. I have a marketing planner, a printed copy of the blue book, and maybe a copy of *Business Kamasutra*. Or maybe just a chapter or two from *Business Kamasutra*." I might offer a book from Dan Kennedy, or whatever gift that I wish to give you. I'll suggest that if you give me your name and address we will ship you these gifts. Another way to do that is to say, "Oh, I'll ship you all those gifts, but just cover the cost of shipping and handling, and pay $5." It's called a self-liquidating offer. The next step is a 20-minute consultation where you are going to meet with me by phone or Zoom. I get to meet you. I get to ask you who you are, what you are, and why you are. In my mind, I want to figure out if we are meant to be together and whether or not I would like you to be on my Ascension Ladder. I will make sure that I will be successful. I'll make certain that I have the capacity to please you in the manner that you want to be pleased. I'm also going to ensure that you are not going to annoy me. You are going to follow my process, and you are going to allow me to practice my trade. When we need to schedule an appointment, I have an automated process that contacts the client and has them schedule the meeting at a time that is easy for them. When it's time for a client to ascend to the next rung on the ladder, that marketing campaign fires off and we start asking them to ascend. (This is a specifically timed process that depends on where they are now, how much business they have done with us, how far along with deliverables we are for them, etc.) By the time you consume my four gifts, I would have installed some software codes in your website and your life, which will give me a better look at your technical landscape. Then, I will most likely have the ability to understand your win condition, what makes you pleased, and if I have the capacity to please you that way. Again, keep in mind that all the communication in this process is automated. Everything bringing you to the 20-minute call is automated.

Now that we are at that stage, we have an initial offer of a free campaign. This is where we would look for a market segment and put together a magnetic offer to attract buyers for you. Maybe it's a free report, a digital gift, a physical gift, and—if you have Keap/Infusionsoft—a three- to five-step email campaign. We do all of that for you—for free even though it costs me $800 to actually do this free gift. Why would I do that? Well, if you do not buy when we meet for the first time, we will have to chase you. Chasing you means we will have to call you; we will have to meet you; we will have to put you on a long-term nurture campaign. All of that takes time, energy, resources, and money. If I spend all that money that I have earmarked to convert an opportunity into a deal to do something for you, most people will accept that offer easily. This gives me a chance to demonstrate that my process will in fact work for you, and that I actually do have a capacity to make you money. We call this "Experience the Genius." We are using our CRM in interaction with you, and using your CRM as a tool in our demonstration.

Think about applying this to whatever you do. How can you automate a series of contacts, automate collection of data useful in determining if someone is an appropriate customer, automate pre-screening so you or your salespeople don't waste their time, and create a demonstration experience? Dan Kennedy has helped many clients with this, including High Point University (hpu.edu), Dr. Phelps' Freedom Founders investment club for dentists (FreedomFounders.com), and Ted Oakley's Oxbow Advisors wealth management (OxbowAdvisors.com), to name a few. Looking in on any of them can give you insight. Of course, you can see mine starting at eLaunchers.com.

CHAPTER 7

Choosing Your Robot Partner: What Marketing Automation Software Do You Want?

By Parthiv Shah

Once you standardize and systematize your sales lead generation, marketing, and sales lead conversion processes, you can automate it, often using off-the-shelf marketing automation tools. There is a dizzying array of marketing automation tools on the market. More arriving almost daily. Personally in my business I use ClickFunnels, Keap / Infusionsoft, and HubSpot more than others. I have clients who use Systeme. io, HighLevel, Ontraport, and others, so it's important that my people are well-versed in multiple software solutions.

I have been an *agnostic* application developer for many years. I have experience working with CRMs like Microsoft Dynamics, Zoho, Act!, Sage, Salesforce.com, TeleMagic, GoldMine, and some homegrown systems in Microsoft Access and FileMaker Pro. For email marketing, I have worked on ExactTarget, VerticalResponse,

Constant Contact, MailerMailer, Gold Lasso, Mailchimp, and a few others. For e-commerce, I have used too many applications to mention. I have been building landing pages using HTML and SQL databases, and we have our own homegrown PURL engine that can generate personalized URLs. We have even created some advanced applications using my PURL engine and named them PURLIZED GURL and GURLIZED PURL. I am also a certified developer on multiple application platforms, and I have affiliate/reseller arrangements with almost all platforms.

So, given all this experience, what do I suggest you need to look for in a CRM?

A few factors:

1. I always recommend the least expensive application that does everything you need it to do. That would include what you expect you'll need in the next 12–24 months. For the sake of ease, you want it to be not only your marketing automation engine, but also your primary data warehouse and data intelligence system. We should always look for off-the-shelf solutions. Sure, you can build a software from scratch with similar functionality, but wouldn't that be over-engineering? Slow? Costly?
2. You'll want a SINGLE DATABASE with multifaceted functionalities. It should create a record for each human in your system, connect your humans to the company they belong to, and keep track of all touch points, all emotional interactions, and all fiscal interactions. You want a SINGLE

DATABASE with CRM functionality, marketing automation functionality, lead capture functionality, e-commerce functionality, Salesforce automation functionality, and an email marketing system *all-in-one*.

3. You want a strong support team and network of consultants that are well-versed in the CRM. A community of consultants is one large collaborative ecosystem. The CRM should support this community and ideally hold it to really high standards, including an initial exam, monthly continuing education credit requirements, and an annual recertification exam just to make sure you are technically competent to belong in the community. In other words, consider all support that exists behind the software.
4. You want an Open API with an ability to push/pull data from almost any ecosystem. The CRM you choose should have a record layout that is adaptable to most business situations.
5. It should have an intuitive and user-friendly interface.
6. It's great if you can find a precedent, meaning someone who is very successful in your industry who has used a particular software—and find out what they used. The culture of sharing information and tribal marketing goes beyond just consultants. I routinely introduce my clients to one another, so they can collaborate and learn from one another.
7. It should integrate ALL channels of communications: email, letter, fax, voice broadcast, text messaging in an online *and offline* environment.
8. It should tell humans what to do, and if humans do not do what they need to do, the system will keep reminding them until the human clicks on a button that "the call was made" or "the gift was sent out."

Here is the Business Kamasutra campaign diagram. You CAN implement this yourself. As a matter of fact, if you reach out to my office and ask me for the campaign blueprint from my Infusionsoft, I will cheerfully give you the campaign blueprint, so you can implement the workflow yourself or hire a consultant to implement it for you.

Figure 7.1: Business Kamasutra Campaign Diagram

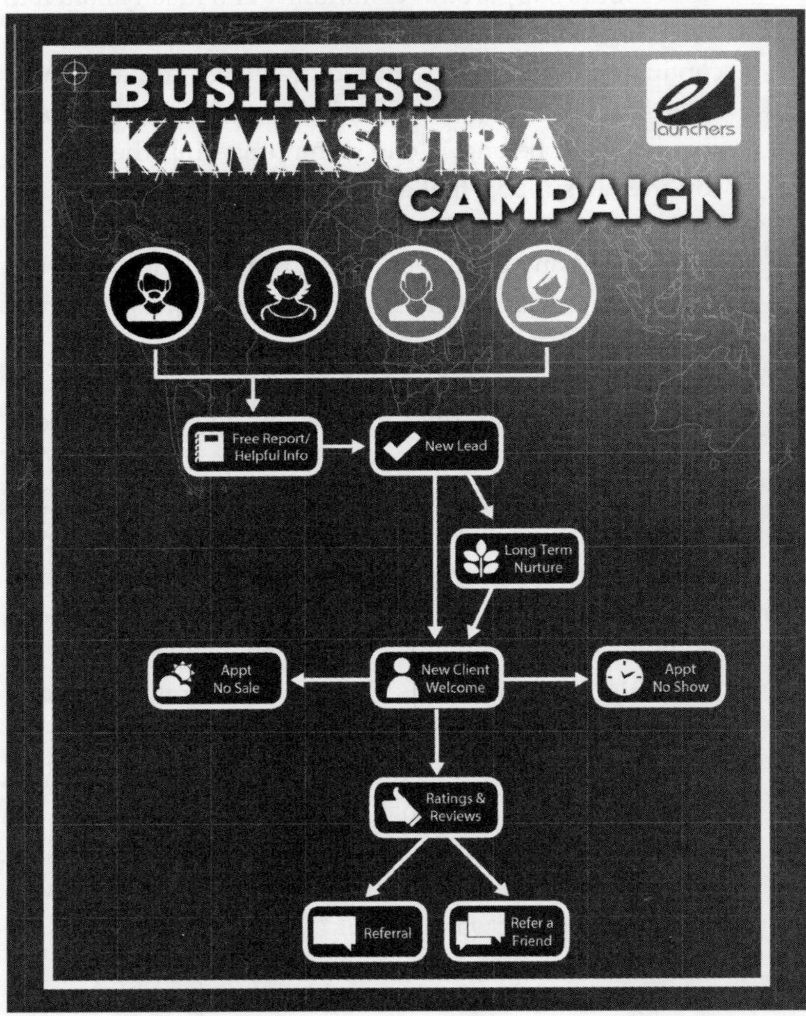

I will even provide guidance and advice to your selected consultant if invited to participate. People think it's strange how free I am with sharing this information; however, I have always thought that when I am not making money, I am making friends, and I need them both.

The Business Kamasutra

I use what I call the Business Kamasutra.

Let us look at various elements of a typical Business Kamasutra campaign and how it works.

1. **Lead Generation:** You will go through the segmentation and approach to generate QUALITY traffic to your website home page, a destination page, or a campaign-specific microsite or landing page. You can use WordPress to build your website and Leadpages.net or an Optimize Press plugin for WordPress to build landing pages, thank-you pages, and tell-a-friend pages. The Business Kamasutra campaign has TWO web forms. One is called Initial Squeeze. This squeeze will capture NAME, EMAIL, and TELEPHONE NUMBER.

   When you fill out your name, email, and telephone number, you are taken to a thank-you page, which is in fact a second squeeze. On the second squeeze page I will deliver the gift I promised on page one, and I will ask you to share your physical address so I can ship you a "box." The box would have a multi-piece literature package we call a Shock & Awe package, a gift of some sort, a book (preferably written by the marketer), and an invitation to take the next step (the low resistance offer).

   There is a school of thought that believes that you should only capture an email, or name and email. Many believe that asking for a telephone number in stage one will reduce your traffic to lead capture ratio. For what it's worth, I agree that asking for a telephone number and making it a required field will eliminate prospects who would have been okay giving you just their email. Personally, I am okay with not talking to prospects who are so concerned about their privacy that I cannot get their contact information. My GOAL is to have a conversation with each prospect, identify their needs, and see if they qualify to be a client. Their privacy preferences are of little relevance to me. I just want to find out who I can talk to and whether or not they can buy! Once I am convinced that I want you as a customer, of course your preferences and priorities are of significant importance to me. (Please understand that this is an opinion, not an insight. I do not claim to have data to substantiate that my position is better. I just have a VERY FIRM opinion on this matter.) The two-step, two-squeeze page method is designed to address the concern of people resistant to disclosing full contact information.

2. **Lead Conversion:** New Lead Follow-Up Sequence is included in this step. This is a three-to-five email follow-up sequence with at least two telephone follow-up steps. The purpose of this sequence is to reiterate the message of the landing page and persuade the prospect to schedule an appointment with the marketer to discuss your mutual interests. If the prospect does not schedule an appointment, the prospect is placed on a LONG-TERM NURTURE sequence. You might omit phone follow-up. You might add follow-up by mail or even FedEx.
3. **Long-Term Nurture:** This is a 24-month campaign that includes a quarterly print magazine, a monthly newsletter

(no newsletter in the month you publish the magazine), a monthly holiday card in print, a twice-a-year gift of some sort, and a twice-a-year email newsletter. This is IN ADDITION to occasional special marketing efforts to convert an unconverted lead and lost opportunity. Such a thing would be impractical without full automation.

4. **Appointment Confirmation Sequence:** When someone clicks to request an appointment, you want to send out information about the initial consultation, directions to your office (if it is an in-person appointment), rules of engagement, what you are trying to accomplish during the initial consultation, and what decision you expect them to make when they meet you for the first time.
5. **Appointment Prognosis Sequences:** There are only three prognoses of an initial sales appointment. There will be either an Appointment No-Show, an Appointment No-Sale, or Sold and "Welcome to Our Practice." You want to have an email, print, and telephone sequence for each scenario, and the correct sequence should be triggered, depending on the prognosis of the appointment.
6. **Ascension and Tell-a-Friend Effort:** When the client makes the initial purchase, WHILE YOU ARE FULFILLING WHAT THEY BOUGHT, you want to ascend them into something else. If you don't sell them something else before the first transaction is over, you are dealing with a "lost customer reactivation situation," which is much more difficult compared to ascending an existing customer. Be prepared to have your next gig ready before you are even close to delivering the first product. Remember, a buyer is a buyer is a buyer is a buyer.

7. **Customer Long-Term Nurture:** This is important. Your customers should also get your long-term nurture. Your name should be in front of your prospects and your clients in a meaningful way.

This describes my Kamasutra for my businesses and ones for clients such as dentists, cosmetic surgeons, law offices, and financial advisors. It may need to be tweaked for your business, but all businesses are more the same than they are different. This campaign will require development of about a dozen web pages, copy for about three dozen email messages, telephone scripts, and a substantial amount of print material.

You may not be able to afford to spend the same kind of money these businesses can to nurture a prospect. If your revenue model does not justify printing a $40 Shock & Awe package, along with a $10 book, in a $20 FedEx box, etc., you may have to settle for just sending a PRINTED LETTER inviting them to a URL where they can read your Shock & Awe online. Do NOT send an email asking them to click. You can use a software like Issuu.com to publish a digital publication and place it right on your website. Visit DentalGrowthMachine.com to read ALL the publications I have published for my business and for practices.

Dan Kennedy Says

Interspecies marriage is a marketing power secret. Nobody should be overemphasizing online and neglecting offline or vice versa. Full integration of all marketing media is key. In this and prior chapters, Parthiv has described systems and software solutions that accommodate both worlds: digital and print/mail. Furthermore, leads might be generated on TV, radio, at trade shows, at consumer shows, by outbound telemarketing, and by other means. They all must be brought into a funnel. All welcomed and managed by the same CRM or bigger system including a CRM.

In this chapter, Parthiv has specifically discussed *print* media. He has mailed over a billion pieces of direct mail for his companies and his clients. He produces print Shock & Awe packages, books, consumer guides. For my clients I do all that and more, often including multiple videos, full-length infomercial format videos, and audio material. While much of it is used in digital forms online, it is also delivered in physical forms: print, DVD, CD, thumb drive. This requires investment in your leads and your customers. If they are not worthy of such investment, I advise getting better leads and customers. If your business's economics prohibit such investment, I urge fixing those economics. I do *not* advise surrendering to only being able to do the cheapest things any more than I would advise only doing what you can do manually.

The great management guru Peter Drucker cautioned against efficiency at the expense of effectiveness. It was and is a brilliant caution. With your Marketing Automation, you are attempting to–and you actually can–defy Drucker's gravity! You are going to improve efficiency and cost-efficiency and effectiveness in concert.

CHAPTER 8

Do You Surrender Too Soon? Too Easily? Because You Have No Choice?

By Parthiv Shah

Wouldn't it be great if you could make a sale after just one well-done sales presentation? Typically, you or your salesperson puts all their energy into doing a great job of the initial sales presentation and then maybe follows up with one phone call or email. In most cases, fewer buy than don't buy, immediately. More money walks away than gets collected. The prospect doesn't buy, and your salesperson believes they are not interested in your product or service and gives up on them. Or your salesperson is blessed with new leads—so why struggle with the recalcitrant? Thanks to your marketing, there may be so much low-hanging fruit, he won't climb trees.

This is how most businesses lose *a lot* of money. It's one of the chief ways that I step in and "find" a lot of money.

According to a study by Brevet, a leading sales consulting firm, 80 percent of sales require five follow-up calls after the first meeting. This seems to be a very consistent fact, supported by many surveys, studies of companies' actual sales results, and work by top consulting firms. But 40 to 50 percent of sales reps give up after one follow-up. Ninety percent of sales pros give up after the fourth call, but 80 percent of prospects say no four times before saying yes. If your takeaway from these stats is that most customers need well-timed and frequent encouragement to say yes by the fifth prompting—you're right. Further, most marketing also has a long tail. Potential buyers first brought to sales situations where they did not buy reappear 6 months to several years later, ready. A lot of business owners and professionals know this happens, yet they don't do anything about it. Dan Kennedy teaches that whatever happens randomly or accidentally can be made to happen methodically, on purpose.

This is why a thorough, patient, and persistent follow-up system is a wise investment. And it can be automated, so it climbs the tree, it struggles with the recalcitrant, it is unaffected emotionally, it has persistence programmed in. My follow-up system uses seven keys:

1. Variety in Media

It's important to remember that different people prefer and respond to different follow-up methods and media. You CAN'T use only what you and your like-minded peers, staff, or friends relate to. Effective sales media includes emails, LinkedIn, or other social media platforms, along with direct mail. Some people process ideas best by reading, others by listening, others by seeing. My follow-up mixes a variety of media over its term.

2. *Timing*

Buying decisions take time, and you don't want to annoy your prospect with too many follow-ups delivered too soon. The higher the price tag, and the more people involved in the buying decision, the longer it may take to get results. On the other hand, you don't want to be too slow or so infrequent that you sacrifice impact and created urgency. How often should you follow up? That depends on your prospect, product or service, timing sensitivity of the purchase, and other variables. In my own case, I do a lot of fairly intense follow-up in a sprint, followed, if needed, by a more gentle, less frequent marathon of follow-up.

3. *Create Value*

Delivering value not requiring purchase in concert with follow-up asking for an opportunity to deliver services or products is a very good strategy. The chiropractor following up might include the "7 Best Ways to Avoid or Lessen the Flu This Flu Season" or "Reminders About Safer, Injury-Free Weekend Sports" along with his offer of exams. Dan Kennedy talks about this in his Magnetic Marketing System® as being a Welcome Guest instead of an Annoying Pest.

4. *Be Interesting and Relevant*

The best follow-up is segmented follow-up. If Bob didn't buy chiefly because of price or affordability, but Charlie didn't buy because of past disappointments with products he thinks of as similar, there is no one-message-fits-all follow-up that can have maximum impact with both of them. They have two entirely different chief interests. You bore them both having to cover both reasons for not buying. This, again, is where your CRM fits. You might have five different lists to choose from, and to put each non-buyer into after the failed sales appointment, each list tied to a different chief objection or concern. Then, five different follow-up sequences of emails, letters, online video, etc.

5. Data-Driven

The best automated systems track who is visiting your site and what interests them. Who views which documents or videos and how much time they spend on each one. A quiz or survey each person completes online can harvest a lot of information useful for the first selling attempt and for follow-up. Offering a choice of two of my six specific free reports can segment potential buyers. Notes kept by salespeople from conversations. We get and use a lot of such data. Some clients use more, some less.

6. Data-Driven Email Automation

With our system, your email sequences (follow-ups) automate, and you can choose your best emails with the highest response rates. You can track open rates and response rates from different subject lines, different days of the week, and different hours of the day to determine and fine-tune the best times to send emails and the best copy to use in those emails to get your desired action.

7. Live Chat & Calling

While your prospect is interacting with your site, you can get their attention and engage them instantly with live chat, which builds instant rapport and closes sales faster. With our unique system, calls can be scheduled, logged into your CRM, and recorded with just one click.

Dan Kennedy Says

Follow-up matters.

In my earliest business life, as a B2B territory rep, I averaged six to seven sales calls each day, roaming about five states. Every night in my motel room, I handwrote Thank-You Notes to the NON-buyers, thanking them for their time and consideration, and enclosing a business card. Fifty percent of non-buyers called me within days to say they had reconsidered and would try the product line in their stores—*because nobody had ever shown them such respect.*

As a consultant and a copywriter, I have developed a 3-step to 16-step Appointment, No-Sale letter sequences for over 50 different types of businesses, averaging a 15 to 30 percent response. One such sequence for financial advisors, applied to attendees of their free workshops who failed to immediately book appointments, using multi-media (online AND offline), up to 18 steps, averaged an 18 percent success rate. In their businesses, the average client was worth $8,000+ in fees and commissions the first year. If they had 100 at their meeting, and they were top performers, they'd book appointments with half, leaving 50 they failed with. My follow-up system for all 18 steps cost $200 per person to implement. 50 x $200 = $10,000. It produced 18 percent of 50 = 9 x $8,000 = $72,000. And just how many times a month would you like to swap $10,000 for $72,000? They could do it two to three times = $62,000 x 2, $124,000 or 3, $186,000, x 10 months a year (skipping July and December) = $1.2 to $1.8 million EXTRA, ADDITIONAL INCOME from the same leads, same meetings. Slow down. Study it. Now here is the "dirty little secret": If we handed them everything and left it up to them and their staffs, NONE OF THEM could ever have implemented all 18 steps properly and on time. Only by fully automating everything for them did this work. Could this work. All they had to do was enter the contact information of those who didn't book appointments the morning after a meeting, click the "GO" button, and they were done.

What marketing and follow-up "machines" do you own, that you just plug in contact info, hit the "GO" button, and you're done?

CHAPTER 9

How I Doubled My Practice 5 Times in 8 Years with Full Marketing Automation

A Case History

by Dr. Carlo Biasucci

I was able to double my practice 5 times over the past 8 years. I've also created the Elite Dental Marketing Program for use by other dentists I coach and consult with.

I want you to imagine you could spend as much time as you want with every single patient before the start of treatment, while they're in treatment, as they're approaching the end of treatment, and post treatment. What would you say to them? How would you say it to them? What would happen with the relationship between you and your patient? As a practicing dentist, you know that's simply not possible when you're both a chief producer and overseer of the business. So how carefully are these conversations being handled right now in your practice? Is the relationship your patients have with you and your practice the best it could be?

I have automated *this.* You would need to hire two to three full-time people to do what this system does automatically. In our internal marketing machine, we have choreographed all the conversations that you could have with the patient and put them in one seamless marketing system with 23 interlocking Infusionsoft/Keap soft campaigns.

What is Infusionsoft/Keap? Infusionsoft is a CRM, a customer relationship management system, and one of the most comprehensive and complex marketing automation systems available to small business, affordable for small business. It ensures that your ideal choreography is carried out exactly when and how you designed it. For example, let's say you generate interest in a particular cosmetic dentistry procedure, and a prospective patient downloads your free report. Your campaign would then go on to send, at the time you specify, emails, direct mail, even physical packages to them, all automatically without you ever having to touch it. The relationship with the prospective patient will be nurtured until they become a patient. Even more importantly, the relationship will continue to be automatically nurtured after a prospective patient becomes a patient of your practice to build trust and rapport.

Most dentists only contact their patients two or three times a year when they're due for the recall visits. Compare that to the dentist who's sending regular weekly emails and direct mail. The dentist with the automated campaign can handle all of this easily and have 60 or more touches in the same time frame. By becoming this familiar and delivering valuable content, your patients and prospective patients will trust you more. They will know you, they will like you, they will be more relaxed when they come in, and they are more likely to follow your instructions, accept the treatment that you're presenting, keep the appointments that they're supposed to keep, be more loyal to your practice, and be the patient you want them to be. And, best of all, unlike two to three employees hired to do all of this, Infusionsoft never gets sick, never quits, and never has a bad day.

The Dynamic Advantage: How to Impress Prospective Patients (Customers) Like No One Else Does

Every business attempts to favorably impress potential customers and customers. If you fully automate your marketing as I have, unlike your competition, you can deliver customized, personal, helpful content and sales messages to your prospects and patients based on who they are, what they need, and how they behave. This is about having a system that increases conversion, which means that you don't spend any more money on advertising than you're currently spending, but you get better results. Before I share this with you, there are three things I want you to understand, and everything I'm going to show you is based on these three things.

The first is to gather information. This one should be obvious. Just like in our practice, you can't prescribe unless you diagnose. In marketing, we want to ask questions to determine if prospects are a right fit. Not every patient is a great patient, right? I'm sure you have patients that you like and patients that you'd rather not have more of. One of the first things the eLaunchers team does with our clients is analyze patient data. This helps us know the characteristics of their best patients. This is true for my dental practice, other practices, and many other types of businesses.

Number two is to customize messaging. The rules have changed. We cannot treat every prospective patient the same. People don't have a lot of time. And if you're sending information that's not relevant to them, they will tune you out and possibly unsubscribe. The good news is that we can send relevant information that's going to help them specifically. This was a big factor in my 5X-in-8-years result: being able to customize messaging.

The third concept that many dental practices aren't consistent with is follow-up. What I mean by follow-up is that every prospect, patient, and lost patient is in a campaign of some sort. This doesn't necessarily mean we're sending stuff every day or every week, but it means we have a plan for them in our marketing system.

I am often asked to simplify and summarize how I 2X'd my practice 5 times in 8 years. The simplest answer is simple. I did 10 times more with each lead, each potential patient, and each patient than I had ever done before, in 1/5 the time—possible only through automation.

More Patient (Customer) Engagement Equals More Case (Offer) Acceptance and More Referrals

There are seven key points of information we have in our system for maximum patient engagement:

Key#1: Last visit activity, or in other words, recency. This is the last time somebody came into your office.

Key #2: How many times somebody visited your office, or frequency.

Key #3: Monetary spend or monetary value.

Key #4: Accepted treatment.

Key #5: If they refused any treatment plan for them.

Key #6: If they have given you any referrals.

Key #7: Rating or reviewing us for others to see.

These are measurements of how well we are doing and of how well each patient is succeeding at behaving as we would like them to behave. I drove my 5X by ending the practice of relying on opinions or feelings—*Oh, we do very well with referrals. Oh, I think our patient value is quite high*. I replaced all that with hard facts. Hard facts can be managed and made better.

I get a daily report with a lot of data, automatically collected and provided. I see who and how many are opening different emails, filling out forms, opening more emails, clicking links, all sorts of things. Staff didn't have to do anything for these emails and clicks and opt-in and web forms. This monitors all the different processes and campaigns.

Let's say somebody sets an appointment. They've gone through part of this. They set an appointment, and they're ready to come in, but they don't show. So now they're a no-show. What do you do? Most offices don't have a formal follow-up process. It's usually maybe a phone call, maybe two phone calls, maybe an email. So what I recommend first in this process is: Let's use more media. Let's use email. Let's use text. Let's send them a letter or two. Let's *really* try to get them back into the office for an appointment. And let's make sure it all happens the same way every time.

The next sequence is the new patient sequence. In this sequence, we have a specific number of communications to talk about different subjects and establish the relationship with the patient. The first email is the welcome email. The second step of the new patient sequence is a direct mail piece. This is a thank-you note that we send out expressing our gratitude to them. After that is Email #2: "Hey, now that we've established a relationship, we're in this together. It's my job as a doctor to keep your mouth healthy, keep your teeth straight, make sure that you don't have any major mouth disasters, and so on." Email #3 says: "Here are the guidelines and the boundaries in which we work. We need to set appointments, you need to keep them, you need to be on time, you need to let us know how we can do better. And if we do all this, we *expect* referrals to your friends and family." Email #4: "We share resources. Here are some resources we have available. Now that you're a member of our family, now that you're a patient of ours, here's our website where we have online resources." This continues.

The next thing I think about is the rating sequence. Usually we want to ask people for ratings and reviews a couple times a year. People usually interact with the dental practice at least twice a year if they're coming in when they should. We use the NPS survey. If you haven't heard of this, this is called Net Promoter Score, and it's one question. We ask them, what is the likelihood you would refer a friend or family member to us? If someone rates us low, we obviously want to follow up and make sure we can maybe better that situation or redeem ourselves. But if someone says, maybe a 9 or 10 likelihood that they would refer a friend or family member, that's great. Then what we do is something really cool. If someone rates us high, we'll know exactly what email address they have. Now everybody wants Google reviews. If you have a Gmail address, we invite you to review us on Google. If somebody has a Yahoo address or a Bing address or Hotmail or whatever, we can invite them to that specific platform in order to review us. And then we get more reviews in those places. And we're not asking people that don't have a Gmail address to review us on Google because that's pretty unlikely and a little difficult for them. We will also direct referral stimulation campaigns to those ranking us 7–10.

We also have a lost patient reactivation sequence. We make this happen automatically. If someone hits a year and they haven't been here, we're going to start this campaign automatically. There's a treatment-not-accepted follow-up sequence. Some of that stretches out for 6 to 12 months.

The more different, specific sequences we built and automated, the more the practice grew.

If you want to see more about everything we do with dental practice growth systems, you can visit: elitepracticeformula.com.

CHAPTER 10

Solving the Case of the Very Messy Marketing Kitchen & More about How to Pick the Best Software for You

By Parthiv Shah

Here's more of the Dr. Biasucci story.

Dr. Carlo Biasucci is an extremely successful dentist and the founder of Elite Practice, Canada's largest training provider for dental teams. Many years ago, he came to me for help with a specific part of his business. We've enjoyed a multi-year, multi-six-figure relationship ever since. Dr. Carlo is one of the smartest men I know. He built a wildly successful dental practice. Sold it. Started a dental coaching business. And became a wildly successful dental coach. He understands his business exceptionally well.

He knows when he needs help. He looks for and hires only the best. He was my first client to hire Dan Kennedy to write

Shock & Awe packages and sales letters. He did this back in 2017. He bought ClickFunnels and Infusionsoft the same year on my recommendation.

Still, we were only part of his team back then. We never implemented a CRM (customer relationship management) or a CMS (customer success management) system for him. He hired someone else to build his website. His blog strategy was all over the place. His tracking tools were homegrown. This led to lost opportunities and unsealed deals. If this weren't bad enough, he hired a sales automation firm that brought in Salesforce.com. This turned out to be a huge mistake. His business processes, team, and customer life cycle were not sophisticated enough to justify an enterprise-level solution like Salesforce. This meant his experience was grossly underwhelming. He never harnessed the true power of Salesforce.com. It was never connected to the rest of his business. His core problems were never solved. Ultimately, he wanted out. He came to me with a problem. He was suffering from a multi-system technological nightmare. He was trying to do too many things all at the same time. This meant his time and resources were spread too thin. His assets were all over the place. He had many brilliant, moneymaking ideas. But none were being properly implemented. He knew things were not going well. He was like a cook in a messy kitchen with no place for his pots. No place for his ingredients. And soup all over the walls.

If you've ever had a successful business seem to spin out of control like this, if you've ever found yourself with a very UN-integrated operation, if you've ever found yourself exhausted and frustrated by being unable to implement your vision, you know "the messy kitchen" problems can happen to anyone. To very smart people. Sometimes, for brief periods of time, this is unavoidable. In fact, Dan Kennedy says that success is often cooked up in a messy kitchen. True. But it can't be sustained in one.

Dr. Biasucci asked me to come in and take a look. If you've ever cooked a pot of soup, you know the soup is only valuable if it's in the pot, in a bowl, or in your belly. Spilled soup, soup on the counter, soup on the stove, or soup on the wall is just a wasted mess. What I saw was like a kitchen nightmare. The soup was everywhere. That is to say, there was systemic data loss, with lead-generating data leaking out of every corner of his business. Dr. Carlo and I had a long conversation. We discussed the importance of keeping a tidy kitchen with all your soup in the pot. He got the picture.

Dr. Carlo asked, "So, Parthiv, what do you know about HubSpot?"

"Quite a lot," I thought, but didn't say.

I could have told him that HubSpot is ideal for anyone looking for an all-in-one inbound marketing, sales, and customer service platform. I could have said it helps you attract clients, convert leads, and close customers with a focus on inbound marketing practices. I could have gone on about its Marketing Hub, Sales Hub, Service Hub, and CRM integration. I could have kept him on the phone for an hour, chatting him up on all that HubSpot can accomplish. I did none of that. I told him I could help. I knew exactly what to do.

HubSpot has a bridge that connects to Salesforce.com. Migrating his underwhelming installation of Salesforce.com and all his data into one new, functional system was doable. We went to work, to build his HubSpot portal, so all his digital assets would be in one place. One platform. Connected to one data set. And instead of all the wasted potential he experienced with Salesforce.com, he would use every millimeter of HubSpot. This means he got more out of his investment. He has a platform that fit his needs. With all his digital assets in one spot. His kitchen was now clean. All the soup stays in the pot. No more systemic data loss. Every lead, every opportunity, and every client got adequate attention in an automated environment.

He went from multi-system technology chaos to ONE SYSTEM on ONE PLATFORM. With HubSpot, we can easily manage customer interactions, marketing campaigns, sales activities, and website analytics. There is seamless data flow and collaboration across the various parts of his business. Efficiency is up. Productivity is up. Dr. Carlo can now easily track customer behavior, marketing performance, and sales activities all in one place.

Dr. Carlo has always been an intelligent man. He's focused. Driven. And it shows. He grew from a start-up (when he first hired me to work on a limited part of his marketing and automation efforts) to a multi-seven-figure coaching business. He did this while throwing things all over his kitchen and spilling far too much soup. He did this with systemic data loss seeping out of dozens of holes in his pot. It excites me to think where he will go from here. What his future has in store for him... All of his soon-to-be-realized accomplishments... All the milestones he will hit then blow past in the next 10 years. If he could grow to the level he's at now while juggling a technological nightmare, I expect him to move mountains now that he's operating with HubSpot.

I first heard of HubSpot many years ago. It was on my radar for a long time. I watched it evolve from its inception back in 2006 to the powerhouse it became over the following decade. I held off on purchasing it, content with what I was using at the time. By 2017, I was convinced I was missing out. I made my purchase—a great decision. Soon, I became a HubSpot Partner, and in 2021, I earned HubSpot Platinum Partner Status. I quickly realized that my company, eLaunchers, is a perfect HubSpot installation. We use every inch of the platform. We've demonstrated a considerable ROI on the non-trivial investment of the software, technology crew, implementation crew, copywriters, and designers. We are truly using HubSpot as intended and are very proud of what we've done with it.

Is HubSpot Right for You? Maybe.

Technologies like HubSpot, ClickFunnels, and Keap are evolving. They have massive R&D budgets. This means their capabilities, strengths, weaknesses, and the reasons why they may be the right tool for you change almost quarterly.

HubSpot is an incredible platform, but it may not be right for you. It's not cheap. There's a learning curve. If your business is not large enough, then it's likely overkill. That being said, there are many reasons why HubSpot blows other platforms out of the water. I can speak from experience with this. I use HubSpot. I love HubSpot. And I highly recommend the platform to the right business owner.

With HubSpot, you keep EVERYTHING in one system.

- Every Domain
- CRM Data
- Contacts and Leads
- Marketing Assets
- Email Marketing Campaigns
- Blogs and Website Content
- Social Media Accounts
- Analytics and Reporting
- Sales Activities
- Forms and Landing Pages
- Customer Service Tickets
- Automation Workflows
- Integration Data
- Feedback and Surveys
- E-Commerce Data
- And More…

Like I said…everything.

This is not an exhaustive list. HubSpot is constantly expanding its features and capabilities.

Keeping everything on one system reduces systemic data loss to near zero. At eLaunchers, we keep our websites on HubSpot. This includes our main website, our DentalGrowthMachine.com, Pixel Estate microsite, our Next100K.com pixel estate, and many other funnel properties. Even the funnels we built with ClickFunnels have HubSpot web forms integrated into ClickFunnels. This bypasses the data interface of ClickFunnels and sends all data directly to HubSpot and nowhere else. While we don't use SEO, PPC, or social media advertising (digital marketing or paid traffic) to grow our agency, we have a comprehensive blog strategy. And our blog CTAs convert these into income-producing assets. Best I can tell, this can only be done with HubSpot. There may be another way to do this with another platform, but I haven't found it. And this is how WE handle our blogs at eLaunchers. We write two blogs a week. They are short, one-idea blogs without soap opera sequences. Each stands alone. They are typically around 750-ish words and have an image to draw attention to the article. Our writer creates a blog header and the content of the blog, then adds a compelling image. They write a month's worth of blogs in the last week of the preceding month. This means that February's blogs are written in the last week of January. This keeps them timely and relevant. Next, our tech crew places the blogs on HubSpot and queues them up for publishing twice a week. This is then done automatically. This means on the designated day and time—for us, Wednesdays and Fridays—HubSpot publishes the blog and pushes it to our social media platforms as posts. On top of that, HubSpot automatically sends an email to our entire blog subscription list once a month. This goes to several thousand people. These blogs contain no pitch. No close. Nothing. Just pure education and some good, wholesome fun. At the end, there's typically a soft offer and call to action, basically inviting them to contact us if they want help or have questions. We keep an eye on WHO opens the blog emails and WHO clicks on the blog links to read more.

Our office manager (also our database administrator and business analyst) goes through the ACTIVE READERS and sends a sales letter with Shock & Awe packages via FedEx.

This powerful way of communicating with leads and generating new clients is an important benchmark test we use to determine if converting to HubSpot is right for a new client. We call it our $75K Rule. Unless you're likely to generate $75,000 in net new revenue using HubSpot with the formula I just outlined and a few other HubSpot automation tricks, we do not recommend you buy the platform. This is why many of our clients and friends use Keap instead of HubSpot.

If you'd like to experience our blogs and compare them to your own, I invite you to do the following. Schedule a Zoom meeting with your staff. Go to www.elaunchers.com/blog. Read some of the blogs. Check out the CTA (call to action) on the blog and article pages. See how I use the top right-hand corner (the golden spot) to my advantage. Explain to your team what you like and dislike about my blogs. Ignore the typos. They are there for reasons I won't go into in this book. After this, go to your blog page and explain what you have to your staff. Then, compare the two experiences. If you want to discuss blog strategy, book an appointment with me using the blog CTA.

What about Landing Pages and Funnels?

When we need a simple funnel, we use HubSpot or Keap. We reserve ClickFunnels for more complex funnels that require certain functionalities unique to ClickFunnels. The site www.meetparthiv.com is an excellent example of a simple landing page we built on HubSpot.

We use "cookies" to track someone who comes to one of our landing pages but does not fill out the form. If someone is already on our list and clicks on a blog article and then moves on to a landing page, but leaves without saying hello, we can reach out to them. Here's an example of how this might work. You're on my blog distribution list. You're going through your emails. See my blog. Click to open the email. This blog has a default call-to-action (book a time with me at www.meetparthiv.com). You click the link. You're directed to my appointment page. You look around, but, for whatever reason, decide not to book. **But I know you were there.** I know you're thinking about booking an appointment. I know you hesitated. Now, I get to act. I could use an automated response. In my case, I prefer having my staff reach out to you. You'll get a phone call. You'll get an email. A text. Maybe also FedEx. With HubSpot, I manage my funnel and landing page AND gain valuable insight into who is responding. This allows me to automate my marketing efforts and capture valuable information about prospective clients.

What about Web Forms?

Web forms are precious digital marketing property. You use them to collect user information, generate leads, and communicate with clients and prospects. There are various types of web forms, including:

- Feedback and Survey forms—great for gathering feedback from clients, patients, and patrons about their experience with your business or product. This helps you understand the customer experience and identify areas for improvement.

- Order forms—vital to anyone selling anything online.
- Referral forms—an easy and inexpensive way to drum up business and turn satisfied customers into your biggest fans.

Whatever application you have for a web form, every form starts with a data field. To create a field or a form, you must first create a permanent field property in your core CRM system. Most software limits how many "custom fields" you can have. For example, Keap allows you to create 100 additional custom fields beyond its default ones. Once you hit that 100, you're at capacity. So, if you use Keap, you have to stay within these limits. For many small businesses, this is fine. They will never reach the limits of this program and can generate web forms that meet their needs without missing a beat. For others with unique needs requiring tailored responses, HubSpot has what you need.

HubSpot gives you:

- 1,000 custom fields for contact records.
- 1,000 custom fields for company records.
- 1,000 custom fields for deal records.

This is a very long runway.

With this platform, you can use web forms on funnels in ClickFunnels, web pages, WordPress, or landing pages in HubSpot. The possibilities are virtually endless.

The Objective Is Immediate, Relevant Response to Every Inquiry or Action by a Potential Customer

I hope you are getting the idea of full-throated, multi-media, multi-step, immediate response, and I hope you are thinking about places where this is *not* happening now in your business. The nature and the immediacy of your response builds or harms trust, furthers and increases interest, or loses it. *We live in Instant World*. Amazon delivers the same day. Alexa or Siri answers our questions instantly. We can draw a circle around an object we see on our screen, in a TV show or movie, and our mobile device will instantly take us to merchants who sell it. *We live in Feedback Land*. Uber tells us how many minutes away our car is. We can see on a map where the driver is and monitor his progress. Social media IS feedback. You post, people Like, post and comment, forward your post. You know how many Likes. In business, in direct marketing, split-tests by print, mail, or even broadcast that took weeks to provide results can now be done in a day by driving traffic in rotation to 2, 4, or 40 sites. CEOs and marketers can know things at the speed of Now. And, when somebody is interested in you and your company, they are interested *now*. If you will use Marketing Automation to accelerate the speed of your relevant response(s), you will capture customers and sales that now slip past you, probably unknown to you.

When a lead is captured, it's time for action. You have a fish on the line. You don't just let it dangle there; you reel it in. Maybe you send an email. Make a phone call. Send a text. Ship an info kit. Maybe you do all of the above. With HubSpot, all of this is automated. Even print-on-demand and shipping of your package can be automated. Simply integrate HubSpot with Mailboxpower.com, and voilà! Like magic, it's all done.

(Mailboxpower.com is one of my favorite vendors. You should look them up.)

One of my prime examples is my benchmark campaign, "Business Kamasutra." This campaign consists of digital and tangible assets delivered to prospects and handled at HubSpot. Business Kamasutra assets include:

- **A New Lead Sequence.** This series consists of seven emails delivered every 3 days. The aim is to persuade a prospect to book an appointment.
- **Pre-appointment Sequence.** Two or three emails that remind my prospect of their upcoming appointment. These set the stage for what they can expect during the call. And most importantly, it makes it far less likely that they will cancel or forget about the appointment the day of.
- **Appointment No-Show Sequence.** Four to seven emails. Not everyone follows through. Sometimes, it's simple because life got in the way. This is my chance to say, "Hey, I understand. Things happen. Let's reschedule for a better time."
- **Appointment No-Sale Sequence.** Four to seven emails. Not everyone is instantly ready to buy. The No-Sale sequence offers valuable information they may not have considered. It's a second, third, and seventh chance at making the sale.
- **Reactivation Emails.** These are quick personal notes. "I've been reading my notes from our call, and I have an idea I'd like to run by you. Please book a time we could talk." These are personal. Effective. And easily implemented with HubSpot.
- **Long-Term Nurture Email Sequence.** This is 52 emails delivered once a week for a year. They offer valuable information or insight into something that benefits your reader and ties into your product or service. These are soft sales. They promote top-of-the-mind awareness. Build trust. Establish you as an authority in your field.

- **Welcome Sequence.** This is a great onboarding tool. Set the expectations for what this new client, patient, or customer can expect. Let them know from the beginning what you expect from them. Set the stage so that the rest of the relationship plays out with everyone's expectations met.

HubSpot facilitates all of this. Once a prospect books an appointment, a "trigger" is created in HubSpot automatically, and sales automation begins. This is marketing automation at its finest. Some or all of the above sequences may be "beefed up" with mail, books, or other physical items sent out, and this, too, happens automatically a pre-set number of days after initial contact.

HubSpot's automation technologies are a micro-step below pure magic. But automation is not the only trick up this platform's sleeve. At eLaunchers, we routinely use HubSpot prospecting tools for *non*-automated outreach efforts.

These include:

- **Telemarketing.** This is where we have warm bodies cold-calling and following up on leads. HubSpot facilitates this with lead management and sales content generation that allows even an average salesperson to produce above average results.
- **Manual Social Media Outreach.** We personally make contact with prospects through social media. This personal one-on-one interaction can help build relationships and trust and allow us to customize the sales experience.
- **One-off (Not Bulk) Direct Mail Outreach.** A personalized letter that directly speaks to your prospect's needs, wants, and desires. Delivered in a FedEx envelope and laced with TLC. This is an effective (and in today's digital age, unique) way of standing out from the rest of your competitors and engaging someone in a meaningful way.

- **Snippet Tools**. These are great ways to store and insert commonly used text or information quickly. These time-saving tools also help with consistency in the delivery of your message. Snippets are easy to use. They all start with a #, so all the user needs to do is put in a #KEYWORD, and Snippet fills up their screen with the next steps.
  1. Conversation Snippets can be used by everyone on your sales team so they know what to say to your prospects.
  2. Email and chat snippets quickly generate text when sending something to a prospect.
- **We use Playbooks** to deliver predefined instructions, guidelines, and scripts that guide office staff and the sales team through interactions with prospects and customers.

  Here's how it works. A prospect picks up the phone and calls. A member of our staff answers and engages in a conversation. They have a Playbook in front of them the entire time. Our staff member begins to ask specific Playbook questions. They type the answers. And the Playbook helps them know what to say next. This keeps the conversation going and the sales experience on track. At the end, a detailed call journal is automatically created and kept in the contact's record.

  A staff member then creates a follow-up task to call back, send a FedEx, or send an email. This is then assigned to someone else on the team. Playbooks mean that appointment setters don't need to be physically in your office. They can be anywhere around the world, and with your Playbook, they know exactly what to say. Once they resolve the call, they simply pass instructions to your designated manager, who then packs and ships a sales letter and a Shock & Awe package with the click of a button on HubSpot.

The Customer/Client/Patient Experience Facilitated by HubSpot (or Other Automation)

We have a thorough, choreographed prospect-to-client experience. You should, too. Yours may need to be different from mine, but here is my main client experience:

- **A Strategy Session Is Scheduled.** This is the catchall stage where all deals are set up.
- **Disqualify/Delete.** This places "bad prospects" in a queue for my office manager to read, journal, and delete. Having my office manager review these prevents me or anyone else from accidentally deleting something that shouldn't have been deleted.
- **Strategy Session No-Show.** When someone fails to show up on a call, they go into this stage of the pipeline. Automation triggers tasks for my office manager to reach out to them and persuade them to rebook.
- **Strategy Session Complete.** This stage is for conversations that end without a clear next step. The conversation went well, but not well enough for the prospect to say yes right then. These deals will die on the vine if not tackled appropriately. At eLaunchers, my wife and I review these monthly to verify they are being managed properly or determine what should happen next.
- **Please Book Another Call.** Maybe there is a need for some follow-up. Perhaps the conversation was not resolved within the allotted time. This stage triggers automation for my office manager to reach out to the prospect and persuade them to book a call.
- **Follow-up Discussion Scheduled**. This is the most desirable scenario. To have the prospect show up on the call as intended, get qualified for further conversation, and have a new call booked before they leave.

- Estimate Requested. This stage is for a prospect when it's time to discuss the details of how we will work together. A prospect is placed here to discuss scope and budget at a higher level. Nothing is finalized yet, but it's close.
- Estimate Submitted. This stage is for a prospect who is ready to sign. We create a QUOTE using the HubSpot Quote tool. Then, we use the HubSpot products tool to add line items to the quote and edit line items in the section as needed.
- Losing Interest. When you get a soft no, the prospect is placed in this stage. This is reserved for those on the fence. They are not ready to commit. They are not ready to say no entirely. This is for the "Let me think about it" crowd. Once a month, we review these prospects and decide what to do with them. These deals may be dead already, but will surely die if not tackled directly.
- Closed—Lost. This is for our firm NOs. The ones who ended the call certain they would never be able to afford me. The ones who seem almost hostile at the mere mention of a monthly charge. These prospects are placed in the Closed-Lost Stage…still, they are never deleted.
- Closed—WON! The finish line…at least as the sales process is concerned. The deal moves to this stage when a prospect digitally signs the quote. We create an invoice in our QuickBooks from HubSpot through an integration between the two apps. This takes the HubSpot quote and converts it into a QuickBooks invoice. Then, all financial transactions happen through QuickBooks.
- Support and Service. Customer service and support automation in HubSpot are robust ecosystems. We paid for the professional-level service hub because we want to use certain features of this hub as our business grows. This is to "future-proof" our business and build a runway for larger deals.

We do this because I believe in investing in technologies you may not need right now, but will likely need in the future. This allows you to build infrastructure functionalities BEFORE the deals arrive.

It is important to understand that all of this has "triggers," automated instructions and alerts to my staff, automated follow-up in each stage of the experience, and information for me to personally inspect. Holes are all plugged. No one is left wondering where we are or why they haven't heard from us yet.

Dan Kennedy Says

The core message of this chapter is GOLD. Parthiv very accurately describes the expectations of the potential customer for instant attention and engagement. When "hot," they go "cold" if left waiting for any length of time. What seems like a perfectly reasonable, slight delay to you in responding, providing information, or fulfilling an information request is, to them, an unbearable eternity that puts your trustworthiness and reliability in question. This is why I and many of my clients invest in FedEx rather than USPS delivery of requested reports, info-kits, books. ANYTHING AND EVERYTHING you can do to speed up your responses as well as to make them more personal and relevant to the prospect is essential to do.

Now, a word about Parthiv. As you can tell by now, he is a very deep thinker about "process." An engineer of the highest order. But he understands something vital, that you need to understand, too, when you go down the rabbit holes of technology, automation, and process. If you've ever wondered what it would be like to have some big-name consulting firm like Bain or Booz Allen work with you and your company, and you are too small of a fish to interest them, you aren't missing much. I have seen some of them at work, and it's comedic. Consider this quote from Amrah Darnaby, a senior consultant at Booz Allen Hamilton: "I generate productivity by celebrating the small wins and I am *most* motivated by seeing success within the process and NOT solely on the outcome." This comes from somebody who never signs the front of a check. Sure, you want to take notice of making any piece of anything work. Yipppee. But this is like celebrating the jockey gracefully getting up onto the horse on the first try. Wow. Pop a cork. In the real world, the only purpose of process IS outcomes. An old mentor went further with "Don't tell me about the labor pains. Just show me the baby." Do you really want to pay giant fees to turn somebody "motivated by process" loose in your company? No. And you must be cautious of process for the sake of process. And wary of process providers in love with process, not outcomes. If you can't write it on a bank deposit slip, it doesn't actually matter.

A popular saying in the tech world is "data is the new currency." As I write this, Salesforce.com is running TV ads with that statement in them, rather inexplicably by the actor Matthew McConaughey, dressed up as a cowboy. But it's not true. Data has to be *converted to* currency. So does automation. In and of itself, it is worthless.

All of the processes described here by Parthiv and demonstrated with examples from his own businesses are *potentially* extremely valuable, if diligently, fully implemented in a way and with tools (software) appropriate for your business.

I also want to point out something obvious, yet missed by many. ALL of this doesn't replace humans. It can reduce the quantity needed. It can make them more effective. But let's take note that Parthiv's company and just about any software/automation company you can name–including Salesforce, HubSpot, Keap, etc., have human sales representatives. Online, however well automated, does not replace offline either. Tech companies advertise a lot on TV, in print media, even with direct mail, printed, in envelopes, with stamps. What is being presented to you here is meant to be integrated with diverse media and with human salespeople and other staff.

CHAPTER 11

AI: How to Make ChatGPT Do All the Research You Could Do & Should Do So That You Crank Out Compelling Copy in Record Time

by Preston Scott Bates

In the cutthroat world of sales, marketing, and entrepreneurship, the daunting task of market research looms like a relentless, insurmountable beast. Picture this: A sea of ambitious individuals, eyes glazed over, staring into the abyss of data, trends, and customer profiles. They're seeking the "Holy Grail" of insights that will magically transform their words into weapons of persuasion. Yet, more often than not, they find themselves lost in a labyrinth of information, with no Theseus to guide them out.

Think of the hours spent poring over data, reading reports that feel like they're written in a foreign language. It's a jungle of numbers and trends. Your eyes glaze over, but you press on, knowing that somewhere in this haystack is your needle. But time is money, and every tick of the clock is a reminder of the opportunity cost mounting with every passing second. The hours spent scouring through data, analyzing consumer behavior, and trying to decode the cryptic language of the market often feel like trying to draw blood from a stone. It's a Sisyphean task, pushing the boulder of research uphill, only to watch it roll back down in an endless, maddening cycle.

Many business owners just give up. With their marketing, they fly by the seat of their pants. They have a mountain of data, but can't mine it. So they just move on.

Imagine the agony of crafting what seems like a perfect piece of copy, only to watch it flop, lifeless and unresponsive, in the real world. The pain of this failure is not just a bruise to the ego; it's a direct hit to the wallet. Every misstep, every miscalculated message, is money down the drain, opportunities slipping like sand through fingers.

Whatever you put out to your customers or potential customers in emails, in videos, in all media has to get the right prospect's attention and really connect with his motivations—but how can you be sure of doing that? The quest to understand the customer, to speak their language, and to tap into their deepest desires—not just push out bland, uninteresting content and sales copy. You're not just writing words; you're a miner, digging deep into the psyche of your market. But here's the brutal truth: Most of you are digging with spoons when you need shovels. You scrape the surface, but never strike gold. The real desires, the raw, unspoken needs of your audience, remain buried, untouched.

Worse still, there's the looming specter of irrelevance. The stakes are sky-high. The market is unforgiving. Miss the mark with your copy, and it's not just a campaign that fails; it's a piece of your business dying. It's not just lost sales; it's a slow fade into obscurity. Imagine the pain of watching customers flock to your competitors, seduced by words you could have written, by secrets you could have uncovered.

What if your competitor cracks the code? What if they speak to your audience's deepest desires while you're still guessing? Can you afford to watch as they capture the hearts, minds, and wallets of the very people you're trying to reach? How will you feel when you realize the market has moved on without you?

So ask yourself, business owner, are you really digging deep enough? Are you truly understanding the fears, desires, and dreams of your audience? Or are you just skimming the surface, playing it safe? What will it cost you if you don't find the key to unlock the true potential of your words? What are you willing to do to ensure that it's your voice that resonates, your message that converts, and your business that thrives?

Frameworks Exist for Crafting Prospect-Based Messaging

Enter Eugene Schwartz and his groundbreaking frameworks from *Breakthrough Advertising*.

Schwartz's methods—famous to most pro copywriters—aren't just theories; they are time-tested, efficient solutions to the very problem you're facing. His approach simplifies the chaos of market research, transforming it into a structured, manageable process. By following his steps, you can save time, reduce frustration, and, most importantly, find the key insights to connect deeply with your customers. You can sort the information obtained for you by Marketing Automation into usable material.

One of the core elements of Schwartz's teachings is the concept of the **5 Stages of Customer Awareness**. This framework is crucial because it recognizes that not all customers are the same. They are at different stages in their journey:

1. Unaware
2. Problem-Aware
3. Solution-Aware
4. Product-Aware
5. Most Aware (2+3+4)

Each stage requires a different approach in your copywriting, a tailored message that resonates with where the customer is in their journey. Why is this important? Because a message that hits the mark can mean the difference between a passing glance and a lasting engagement. For instance, a Problem-Aware customer knows they have a problem, but might not know solutions exist, or at least may not have looked for them. Your copy needs to empathize with their pain and introduce your product as a solution. Conversely, a Product-Aware customer already knows what options are out there from your competition and knows (or thinks he knows) what you offer, but has not acted on what he knows to make a purchase.

Using Schwartz's frameworks is not just about making your life easier; it's about being effective and efficient in your marketing efforts. It's about understanding your customers on a deeper level and crafting messages that not only resonate with them but also compel them to take action.

How to Do Meaningful Research to Categorize Your Prospects by the Schwartz Frameworks

Conducting market research for effective copywriting, as outlined by Eugene Schwartz in *Breakthrough Advertising*, involves a systematic and thorough approach that encompasses understanding both the market and the product in-depth. Here are the steps in the process.

Thorough Analysis of the Market

This initial step involves an in-depth exploration of the market for the specific product. The aim is to understand the market's breadth and depth comprehensively. This includes identifying the key emotional forces that drive the market and defining these forces in terms of a singular image, desire, or need. This step is crucial in understanding what motivates the market, what it seeks, and how it behaves. WHY do the people you intend selling to respond to certain analogical companies and their advertising? WHY do they buy comparable products and services—and not buy others? Whether you are a start-up or you have been in business for years, looking for answers to these questions is important to do well repeatedly, not just once.

Meticulous Study of the Product

After analyzing the market, the next step is a detailed examination of the product itself. This involves understanding what the product is, its functionalities, and the benefits it offers. The focus here is on identifying all the satisfactions and performances the product can deliver. The goal is to distill these into a single compelling claim that effectively taps into the strongest emotional force within the market. You aren't just presenting a product and its virtues. You are connecting certain virtues of the product with what you know about the market's motivations; its collective psyche.

Synthesis of Market and Product Insights

Combining the insights from the first two steps leads to the formulation of the core theme of the advertising effort. This theme encompasses the desire the market demands and how the product satisfies it, the need the market feels and the solution the product offers, and the identification the market seeks and how the product expresses it.

Expressing the Theme

With the theme identified, the process moves into expressing it effectively. This involves exploring the market's maturity level and understanding how much the market knows about the product, what they have been informed about similar products, and their overall interest and care in these aspects.

Identifying the Point of Entry for the Headline

From the comprehensive analysis comes the crucial point of entry for the advertising headline. This is the point that garners the greatest interest and acceptance from the target audience. It could be centered around the product's unique features, its price, performance, the satisfaction it promises, the need it fulfills, or a unique aspect of the market itself.

Commencement of Copywriting

Finally, with the point of entry identified, the actual process of writing the copy begins. This marks the transition from research and planning to the creative expression of the core message in a way that resonates with the target audience.

In summary, the process is about deeply understanding both the market and the product, synthesizing these insights to form a compelling advertising theme, and then effectively expressing this theme to resonate with the target audience.

How to Get Them to Tell You Where They Are in the Framework and What They Really Want

There are a lot of information-gathering tactics and tools usable in online media, easily automated. An innocent-looking survey can automatically put everybody saying Yes to Question #3 into one group, to automatically get a specifically relevant email sequence and put everybody saying No into a different group to get a different email sequence. A choice of a particular Special Report out of a dozen offered can reveal a specific interest and trigger its own automated marketing. From the chiropractor, John Prospect picks "How to Stop Backache and Increase Mobility for a Better Golf Game" and Jim Prospect picks "How to Look & Feel 10 Years Younger—Without Surgery or Drugs." Neither should get a generic presentation about chiropractic. Both shouldn't get the same presentation. These are simple tricks you can build into your marketing.

For prospect and customer information gathering, organizing data, classifying customers, and writing copy, I've learned to love "chat." Enter "Chatvatars"—using ChatGPT to create avatars for each of Schwartz's 5 Stages of Customer Awareness.

Are you ready to see what leveraging a tool like chat could look like for your business? Here's the prompt I gave chat for this illustrative example:

For a book called *No B.S. Guide to Successful Marketing Automation: The Ultimate No Holds Barred Guide to Using Technology, Automation, and Artificial Intelligence in Marketing,* write 500 words motivating a frustrated small business owner—who's overwhelmed with the process of researching and writing copy, and who's curious about using ChatGPT to create avatars for each of Eugene Schwartz's 5 Stages of Customer Awareness. Call the process "Chatvatars." And list 17 unique and compelling reasons why the small business owner should use ChatGPT to create the "Chatvatars" that focus on the functional, dimensional, and emotional benefits of saving time, increasing the likelihood of success, saving frustration, saving money, and writing better copy.

And here's what it gave me:

Time Efficiency: ChatGPT can generate these avatars quickly, giving you more time to focus on other business areas.

Consistency in Messaging: By using ChatGPT, you ensure a consistent tone and style across all stages, maintaining brand integrity.

Deeper Customer Insights: ChatGPT can help analyze and interpret customer data, providing deeper insights into each stage of awareness.

Cost-Effective: Outsourcing this task or hiring a team for it can be expensive. ChatGPT is a cost-effective alternative.

Stress Reduction: Reducing the manual labor of research and copywriting eases your mental load.

Enhanced Creativity: ChatGPT can offer creative suggestions that you might not have considered, enhancing your campaign's innovation.

Customization: Tailor your messages for each stage of awareness with personalized suggestions from ChatGPT.

Data-Driven Decisions: ChatGPT's algorithms are based on extensive data, leading to more informed decision-making.

Market Adaptability: Quickly update your avatars as market trends change, keeping your business agile.

Improved Copywriting: With suggestions from ChatGPT, your copy can become more persuasive and engaging.

Error Reduction: ChatGPT can help identify and correct errors in your research or copy, enhancing accuracy.

Scalability: As your business grows, ChatGPT can easily scale up to meet increased demands for content and research.

Competitive Advantage: ChatGPT allows you to stay ahead of competitors by leveraging advanced AI technology in your marketing strategies.

Learning Opportunity: Working with ChatGPT can deepen your understanding of AI and its applications in marketing.

Better Engagement: Tailored content for each awareness stage leads to better customer engagement and conversion rates.

Emotional Resonance: ChatGPT can help craft messages that resonate emotionally with your audience, crucial for effective marketing.

Quantifiable Results: Over time, using ChatGPT for your avatars will yield data that you can analyze to see clear results and ROI.

Now, notice how much stronger the copy is when I enhance it with a prompt derived from two titans of direct response, the late great Clayton Makepeace and the inimitable Todd Brown:

According to Todd Brown's definitions of functional, dimensional, and emotional benefits, give me a functional, dimensional, and emotional benefit for each of these [number of features] features for my [type] brand:

[number of features] Features: [list your features]

First, let me give you clear and detailed explanations of terms according to Todd Brown…

We'll start with a clear explanation of the term "feature" according to Todd Brown:

A "feature": Simply a fact about the products

As a reminder, here are all the features of our products I need unique functional, dimensional, and emotional benefits for:

[list your features]

Next, let's clearly define the term "functional benefit":

A "functional benefit": What a given feature does for the prospect. How a given feature improves the prospect's life. The impact the feature has on the prospect.

State the functional benefit in comparative terms.

For example,

Feature: EVA Antimicrobial Foam

Functional Benefit: Effectively kills 50 percent more bacteria and fungi to keep your shoes hygienic and fresh.

"Keep your shoes hygienic and fresh" is a benefit of the way the "EVA Antimicrobial Foam" feature functions or performs.

Now, let's clearly define the term "dimensional benefit":

The "dimensional benefit," originated by renowned copywriter Clayton Makepeace, describes what the functional benefit looks like in the real life of the prospect. How the functional benefit plays out in their lives.

According to Todd Brown…

The dimensional benefit: Paints a vivid picture of how the prospect experiences the functional benefit in his or her life.

Start the sentence of your dimensional benefit with the phrase "So you can…"

For example,

Functional Benefit: Effectively kills 99.9 percent of bacteria and fungi to keep your shoes hygienic and fresh.

Dimensional Benefit: So you can wear your shoes without worrying about any foul odors or stains caused by bacteria and fungi.

Finally, let's clearly define the term "emotional benefit":

An "emotional benefit": The emotional impact of the dimensional benefit. How the prospect will feel emotionally when they experience that dimensional benefit.

Start the sentence of your functional benefit with the phrase "You will feel…"

For example,

Dimensional Benefit: So you can wear your shoes without worrying about any foul odors or stains caused by bacteria and fungi.

Emotional Benefit: You will feel confident and self-assured knowing that your shoes are always hygienic and fresh, no matter how much you wear them.

Make sense?

So again, I need you to give me a different functional, dimensional, and emotional benefit for each one of the following features of my brand: [list your features]

Now watch how much better that list of 17 reasons gets when written as functional, dimensional, and emotional benefit statements. Here's my prompt:

Now write functional, dimensional, and emotional benefits for each of the 17 compelling reasons listed above.

And here's the output:

Absolutely! Let's break down the functional, dimensional, and emotional benefits for each of the 17 compelling reasons to use ChatGPT for creating Chatvatars:

Time Efficiency

- Functional: Streamlines content creation, significantly reducing time spent on marketing tasks.
- Dimensional: So you can allocate more time to other critical aspects of your business, enhancing overall productivity.
- Emotional: You will feel a sense of relief and achievement, knowing that valuable time is being saved.

Consistency in Messaging

- Functional: Maintains a uniform tone and style across all content, reinforcing brand identity.

- Dimensional: So your brand's message remains clear and consistent in the customer's mind, building trust.
- Emotional: You will feel confident and proud of presenting a cohesive brand image.

Deeper Customer Insights

- Functional: Provides in-depth analysis of customer data for more accurate targeting.
- Dimensional: So you can craft messages that truly resonate with your audience's needs and preferences.
- Emotional: You will feel enlightened and empowered with a deeper understanding of your customers.

Cost-Effective

- Functional: Reduces expenses associated with market research and copywriting.
- Dimensional: So you can invest more resources in other growth areas of your business.
- Emotional: You will feel financially savvy and resourceful, maximizing your budget's impact.

Stress Reduction

- Functional: Eases the burden of marketing tasks, leading to a more manageable workload.
- Dimensional: So you can operate your business with less anxiety and more focus.
- Emotional: You will feel more relaxed and in control, free from overwhelming marketing pressures.

Enhanced Creativity

- Functional: Offers creative suggestions and ideas, sparking innovative marketing approaches.
- Dimensional: So your marketing campaigns stand out with fresh and unique content.

- Emotional: You will feel inspired and inventive, breaking free from conventional marketing tactics.

Customization

- Functional: Tailors content to specific audience segments for more effective communication.
- Dimensional: So you can connect more personally with different customer groups.
- Emotional: You will feel a sense of satisfaction from delivering highly relevant and engaging content.

Data-Driven Decisions

- Functional: Bases marketing strategies on solid data analysis for improved outcomes.
- Dimensional: So your marketing choices are more informed, reducing guesswork and risk.
- Emotional: You will feel more confident and assured, knowing your decisions are data-backed.

Market Adaptability

- Functional: Quickly updates marketing content in response to changing market trends.
- Dimensional: So your business stays relevant and competitive in a dynamic market.
- Emotional: You will feel agile and proactive, capable of navigating market shifts successfully.

Improved Copywriting

- Functional: Enhances the quality and effectiveness of your marketing copy.
- Dimensional: So your messages are more compelling and persuasive to your audience.
- Emotional: You will feel proud and fulfilled, seeing higher engagement and response rates.

Error Reduction

- Functional: Identifies and corrects inaccuracies in content, ensuring high-quality communication.
- Dimensional: So your brand maintains credibility and professionalism in all messaging.
- Emotional: You will feel reassured and trustworthy, presenting error-free and reliable content.

Scalability

- Functional: Adapts easily to increased content needs as your business grows, without loss of quality.
- Dimensional: So you can expand your marketing efforts seamlessly, catering to a broader audience.
- Emotional: You will feel prepared and capable, knowing your marketing can grow with your business.

Competitive Advantage

- Functional: Utilizes advanced AI technology, giving you an edge over competitors.
- Dimensional: So you can lead in your market with innovative marketing strategies.
- Emotional: You will feel a sense of accomplishment and superiority, staying ahead in the competitive landscape.

Learning Opportunity

- Functional: Provides insights into AI applications in marketing, enhancing your knowledge.
- Dimensional: So you can stay informed and ahead in marketing technology trends.
- Emotional: You will feel enlightened and forward-thinking, continuously evolving with technological advancements.

Better Engagement

- Functional: Delivers content that resonates more effectively with the audience, improving engagement.
- Dimensional: So you can build stronger connections with your customers, encouraging loyalty and repeat business.
- Emotional: You will feel a sense of connection and satisfaction from positive customer interactions.

Emotional Resonance

- Functional: Crafts messages that emotionally connect with your audience, deepening their relationship with your brand.
- Dimensional: So your brand can evoke powerful emotions, driving customer action and attachment.
- Emotional: You will feel a profound sense of fulfillment, knowing your brand is making a meaningful impact.

Quantifiable Results

- Functional: Yields measurable outcomes from marketing efforts, allowing for clear analysis of effectiveness.
- Dimensional: So you can continuously refine and optimize your marketing strategies based on solid data.
- Emotional: You will feel a sense of control and assurance, being able to see and understand the direct impact of your marketing decisions.

Each of these benefits highlight how using ChatGPT to create Chatvatars can transform your marketing strategy, not only in functional and practical ways but also in creating a deeper emotional connection and understanding between your brand and your customers.

So now you can see how powerful of a tool Chat can be at simply writing benefit statements with little training on the audience. You want to train your AI writer on your audience *before* assigning it writing prompts, like headlines, so that the generative AI tool understands exactly who you want to speak to. This is the difference between average and good or better than good copy. Here's an example prompt you can use for what I call the "Chatvatars."

Give me five customer avatars for (offer) teaching (target) whose company (describe company) how to (insert outcome).

Create one avatar for each Stage of Awareness as described in Breakthrough Advertising.

These are:

- Unaware
- Problem-Aware
- Solution-Aware
- Product-Aware
- Most Aware

The target market is [market]. The offer's website is [website]. It's a [price positioning] [type of offer] in the [niche] that helps [target market] to [benefit].

Put the avatar in the following format:

1. Avatar Details:
   Stage of Awareness:
   Name:
   Gender:
   Job:
   Annual Salary in USD:
2. Personal Profile:
   Top 5 Favorite Movies:
   Top 5 Favorite Books:
   Top 5 Visited Websites:
   Top 5 Relevant Social Media Influencers:

3. Psychographics
   5 Personality Traits:
   5 Values:
   5 Interests:
4. Pain Profile
   Big Secret Fear:
   Big Complaint from the Existing Solution:
   Top 5 Primary Pain Points:
5. Desire Profile
   Big Secret Desire:
   Big Lifestyle Desire:
   Top 5 Primary Desires:
   Top 5 Emotions/Feelings:
   Top 5 Beliefs/Identifications:
6. Shopping Behavior
   Top 3 Primary Decision Triggers:
   Purchase Frequency:
   Prior Purchases:
   Average Order Value:
   Preferred Channels:
7. Wants
   Wants to Gain:
   Wants to Be:
   Wants to Do:
   Wants to Save:
   Wants to Avoid:
8. Three-Dimensional Profile
   Seeing:
   Thinking:
   Hearing:
   Feeling:
   Saying:
   Doing:

Make sure each avatar is at least 900 words.

Start with the problem-aware avatar.

Then, in part two, you have it write a "Dear Diary" entry as the problem stage, for example, the avatar using emotionally laden diction for two reasons. First, it helps train the generative AI on successive prompts so that you have better copy. Secondly, it helps you as the writer construct arguments with more empathy and insight so your copywriting converts, which is the goal, bucko! So use this prompt next:

Dear Diary: Imagine you are the problem-aware avatar. Taking the psychographics, pain profile, desire profile, wants, and three-dimensional profile into account, write me a diary entry of at least 1,000 words. Use passionate, visceral, emotional language that reveals your deepest feelings, fears, and desires.

Here's an example of just how powerful these tools can be for the aspiring entrepreneur who wants leverage. It's an example of Chatvatar I developed for a client of Parthiv Shah, Ray Miller, who sells into a small Total Addressable Market. Therefore he must increase average transaction size and lifetime value since there's an upward limitation to the number of new clients he can acquire. The point being, his marketing must be very dialed in. It needs to hit. To resonate.

Here's an example of the problem-aware avatar I developed for his commercial real estate furniture installation business:

Avatar 1: Problem-Aware

1. Avatar Details:
    - Stage of Awareness: Problem-Aware
    - Name: Richard Evans
    - Gender: Male
    - Job: VP of Corporate Real Estate
    - Annual Salary in USD: $250,000

2. Personal Profile:
   - Top 5 Favorite Movies: *The Big Short*, *Inception*, *Wall Street*, *The Wolf of Wall Street*, *Moneyball*
   - Top 5 Favorite Books: *Good to Great* by Jim Collins, *The Power Broker* by Robert Caro, *The Innovator's Dilemma* by Clayton Christensen, *Thinking, Fast and Slow* by Daniel Kahneman, *The 7 Habits of Highly Effective People* by Stephen Covey
   - Top 5 Visited Websites: Bloomberg.com, WallStreetJournal .com, LinkedIn.com, HarvardBusinessReview.org, Forbes .com
   - Top 5 Relevant Social Media Influencers: Simon Sinek, Gary Vaynerchuk, Ray Dalio, Arianna Huffington, Tony Robbins

3. Psychographics
   - 5 Personality Traits: Detail-oriented, ambitious, strategic, analytical, pragmatic
   - 5 Values: Efficiency, innovation, reliability, leadership, integrity
   - 5 Interests: Commercial real estate, investment strategies, corporate leadership, technology trends, golf

4. Pain Profile
   - Big Secret Fear: Becoming obsolete in the rapidly changing corporate landscape
   - Big Complaint from the Existing Solution: Inefficiency and lack of integration in current workplace installation solutions

- Top 5 Primary Pain Points: Time-consuming processes, poor communication between vendors, unexpected delays, budget overruns, inadequate customization

5. Desire Profile
   - Big Secret Desire: To be recognized as a visionary leader in corporate real estate
   - Big Lifestyle Desire: A balanced life where professional success and personal time are harmoniously blended
   - Top 5 Primary Desires: Streamlined processes, reliable partnerships, cutting-edge office spaces, cost-effectiveness, industry recognition
   - Top 5 Emotions/Feelings: Frustration, ambition, determination, anxiety, pride
   - Top 5 Beliefs/Identifications: The importance of a well-designed workspace, the value of time, the power of networking, the necessity of innovation, the impact of a strong corporate culture

6. Shopping Behavior
   - Top 3 Primary Decision Triggers: Efficiency improvements, cost savings, testimonials from industry peers
   - Purchase Frequency: As needed for corporate real estate projects
   - Prior Purchases: Various office furniture, technology upgrades, consultancy services
   - Average Order Value: $500,000–$2,000,000
   - Preferred Channels: Direct sales, industry conferences, professional networks

7. Wants
   - Wants to Gain: Recognition as an industry leader
   - Wants to Be: A pioneer in innovative workplace solutions

- Wants to Do: Implement turnkey solutions that revolutionize how spaces are created
- Wants to Save: Time and money for his company
- Wants to Avoid: The stress and inefficiency of traditional installation processes

8. Three-Dimensional Profile
   - Seeing: A vision of streamlined, state-of-the-art offices
   - Thinking: About the logistics, budget, and potential ROI of new installation solutions
   - Hearing: Feedback from other industry leaders and keeping an ear to the ground for innovative solutions
   - Feeling: The pressure to deliver and the excitement of potential success
   - Saying: Advocating for change and improvement in corporate real estate processes
   - Doing: Researching new solutions, negotiating with vendors, planning future projects

Richard is keenly aware of the problems in his industry and is always looking for efficient, reliable solutions that can enhance the value and functionality of the corporate real estate he manages. He's well connected and respected in his field, driven by a desire to succeed and be recognized for his innovative approaches to traditional challenges.

And here's what the "Dear Diary" entry looked like:

Diary Entry: Richard Evans–March 29, 2024

Dear Diary,

It's late, and the city's lights are just blurs beneath my office window, a reminder of the restless energy that drives me. Today was another marathon, a relentless pursuit of excellence that I both revere and dread. My mind is a battlefield, a tumultuous mix of ambition and frustration, and tonight, the silence is almost too loud.

I am haunted by a profound sense of urgency, a gnawing feeling that the ground beneath me is shifting faster than I can adapt. The world of corporate real estate is evolving at a dizzying pace, and my fear of becoming obsolete is a shadow that looms over every decision I make. I am Richard Evans, a man atop a mountain who fears the erosion that could send him tumbling.

As I sit here, the weight of my role as VP of Corporate Real Estate at a Fortune 500 company is a constant companion. My decisions shape the landscape of our corporate environment, and the pressure to innovate, to lead, and to succeed is a relentless force. The industry whispers about efficiency and innovation, but I'm screaming inside for a solution that truly understands the complexity of what we need.

The inefficiency of our current workplace installation solutions is a thorn in my side, a source of continuous irritation. I am tired of the time-consuming processes, the poor communication, the unexpected delays, and the budget overruns. Each project feels like a gamble, a dice roll where the odds are never in my favor. I crave a streamlined process, a symphony of coordination where every piece moves in perfect harmony.

But beyond the irritation and the frustration lies a deeper, more visceral fear. I am haunted by the possibility of failure, the idea that I might choose a path for my company that leads to a dead end. My position demands innovation, but innovation carries the risk of the unknown. The thought of implementing a solution that falls short of our dynamic needs is a nightmare that wakes me in the cold, early hours of the morning.

Yet, for all the fears and frustrations, there burns an unquenchable fire within me, a desire to carve my name into the annals of industry leaders. I long to be the visionary, the trailblazer who revolutionizes the corporate real estate world. I envision spaces that aren't just built, but are birthed from ingenuity and foresight, environments that breathe productivity and inspire creativity.

I dream of streamlined processes and cutting-edge office spaces that reflect the prestige of our company. I want to build legacies, not just buildings. I want to be the one who brings the future into the present, who sees beyond the horizon and brings that vision to life. This isn't just about constructing spaces; it's about crafting the future of how our company and industry operate.

But my desires go beyond the professional. I yearn for a life where my professional success does not come at the cost of personal time, where I can stand at the pinnacle of my career and still enjoy the view. I want to be recognized not just for the spaces I create, but for the innovation and foresight I bring to every aspect of my work.

My emotions are a whirlwind, a mix of frustration and ambition, anxiety and pride. I am driven by a relentless determination to succeed, to prove that my vision is not just a dream, but a blueprint for the future. I am fueled by the belief in the importance of well-designed workspaces, the value of time, and the power of a strong corporate culture.

As I look to the future, I see a path lined with decisions, each a stepping stone toward my ultimate goal. My shopping behavior reflects this journey, with every purchase a careful calculation toward efficiency and innovation. I seek testimonials from industry peers, evidence of success that can guide my way.

I want to gain recognition as an industry leader, to be the pioneer of innovative workplace solutions. I want to implement turnkey solutions that change the game, that make the stress and inefficiency of traditional installation processes a thing of the past. I want to save time and money, not just for my company but for myself, to find that elusive balance between success and serenity.

As I sit here, alone in the quiet, I see my vision for the future. I think about the logistics, the potential ROI, and the impact of every decision. I hear the whispers of industry trends and the feedback from my peers, the collective voice that guides my way. I feel the pressure to deliver, the excitement of potential success, and the satisfaction of a job well done.

I am saying that change is necessary, that we must advocate for improvement and innovation. And tomorrow, I will do what I have always done: research new solutions, negotiate with vendors, and plan for the future. I am Richard Evans, a man aware of the problems that plague us, but more importantly, a man who is seeking, striving, and never yielding in the pursuit of solutions.

Three Reasons to Make AI Your Hardest Working Marketing Researcher and Writer, Right Now

Hopefully you can see how powerful this is for anyone harried to produce results and who can't be bothered to invest years and hundreds of thousands of dollars to become a great copywriter, as I have invested. In today's rapidly evolving digital landscape, small business owners are constantly seeking tools to streamline their marketing efforts. The integration of AI-driven solutions like ChatGPT can revolutionize how small businesses approach market research and copywriting.

For the overwhelmed entrepreneur, here are the top three reasons why adopting ChatGPT is a game-changer for making research faster, easier, more efficient, and for producing copy that converts:

1. Accelerated and Streamlined Market Research

ChatGPT significantly speeds up the market research process by quickly analyzing and synthesizing vast amounts of data, providing actionable insights in a fraction of the time it would take manually. This accelerated research process means you can respond rapidly to market changes and customer needs. Adopted, you can stay consistently ahead of market trends and customer preferences, enabling you to make swift and informed decisions about your product development and marketing strategies. This agility in adapting to market dynamics is crucial for small businesses that need to optimize their resources efficiently.

2. *Enhanced Quality and Relevance of Marketing Copy*

The better and more robust your Marketing Automation is, the more content and sales copy it has to be fed. ChatGPT's advanced algorithms help create highly relevant and engaging marketing copy, tailored to the specific needs and stages of your target audience. This ensures that your marketing messages resonate more deeply with potential customers, leading to higher engagement and conversion rates.

Your marketing campaigns can effectively address the unique concerns and desires of different customer segments, creating a more personalized experience for your audience. This tailored approach helps in building stronger relationships with customers, enhancing brand loyalty and advocacy.

3. *Cost-Effectiveness and Resource Optimization*

By automating and optimizing the research and copywriting process, ChatGPT helps in significantly reducing the costs and resources typically required for these tasks. This is especially beneficial for small businesses with limited budgets and manpower.

Preston Bates is a Scottsdale, Arizona–based entrepreneur and copywriter who owes no small debt of gratitude to Dan Kennedy for rapidly accelerating his direct response chops and to whom he can reasonably attribute hundreds of thousands of dollars in revenue. You can connect with him at @prestonsbates.

Dan Kennedy Says

At minimum, the ChatGPT's work provides raw material, rough drafts, and outlines to make your writing output faster and easier. The whining I often hear from business owners shown how to use all media and automated communication, weekly memos, monthly newsletters, mini-books, etc. is (a) *I can't write,* (b) *I don't have time—I've got teeth to drill, cars to repair, shelves to stock.* Well, now you can, fast. At maximum, at some competitive levels, this "robot's" work is good enough, although I do NOT share Preston's enthusiasm for AI as a substitute for the human writer or copywriter. My friend and colleague the late Zig Ziglar said, "If two of us agree on everything, one of us is unnecessary." At optimum, if you are able to hire high-skill professional copywriters, in-house or freelance, this can help you provide them with a lot more information to work with.

It can be a struggle to "feed the beast." When you set up a complex, sophisticated, complete media platform of your own, to act as a fence around customers and prospects, by communicating frequently and consistently in multiple media, you create a hungry beast that needs a lot of content relevant to your audience. This *can't* be allowed to stop you from creating such a beast. You *must* find ways to empower your beast. I designed my own company to have no less than 262 communication touches in a year. Although I no longer own or manage it, I do continue contributing to it, and it operates at a similar level of frequency of contact with its Members, customers, and interested potential Members. If you take advantage of the Special Offer on Page 183, you'll experience this.

I'll add: There are a lot of sources of content: things to comment on, news of the day specifically relevant to your market, holidays (there are hundreds during the year), your own experiences, books you read. From any of this, you can get the jumping-off-point for an email, a post, a blog, a new sales letter. There is also a lot of material in the public domain that you can use, free. To do this, start with the book *The Public Domain: Find and Use Free Content for Your Website, Book, Newsletter and More* published by Nolo Press. Parthiv did an outstanding job using the famous Russell Conwell speech *Acres of Diamonds,* from public domain, as content in his own mini-books, promoting himself as a data scientist finding your acres of diamonds. If you would like a copy, email Parthiv at pshah@eLaunchers.com.

CHAPTER 12

A Few Thoughts about AI

by Dan Kennedy

Here are two items reprinted from my *Dan Kennedy Letter*, a monthly newsletter published by Lillo/Kennedy, available by subscription by contacting Pete Lillo's office by phone or fax: 330-922-9833.

Will the Robots Love Us? Forever 'n Ever?

IN 1932, Aldous Huxley's seminal novel *Brave New World* was published, meant as a critical, satiric response to the idea of a utopian society free of labor with all equitably fed, housed, and cared for by machines, then promoted by sci-fi author H. G. Wells AND as a statement of Huxley's increasing disenchantment with the changes occurring around him. Huxley's *Brave New World* delivered a nightmarish prediction of a future in which humanity, morality, and opportunity have been erased. Many subsequent novels and movies owe their themes to Huxley—even *Planet of the Apes*. Right now, there's an explosion of noise about letting "AI" write your ads, marketing materials, content, as well as nonfiction and fiction books, do the work for doctors, lawyers, fly airplanes and more. But we should note that the "A" stands for ARTIFICIAL. In 1939, Napoleon Hill wrote a series of how-to articles based on his work in advertising, titled *"How to Write **HUMAN** INTEREST Sales Letters."* AI fakes it. Mimics it. Aggregates it. But it ***can't actually BE* human** so it can never achieve true human-to-human connection. It does its thing mechanically; the high-skill human writer or copywriter does much by human instinct and impulse, by honest, often raw emotion. *AI makes calculations absent morality*. Would AI have tried landing the plane on the Hudson River and saving everybody's lives like Capt. Sully did? No. Watch the NFL now. Amazon's AI calculates the percent likelihood of certain plays right before or as they are happening, and *most games are won by coaches calling and QBs and players making the plays AI said NOT to run*. **Can AI write AS me? No.** It can write *almost like* me. I've been shown samples. It is good enough to fool some people, so is that good enough? If enough people think so, and care less about authenticity or morality (as is increasingly evident), then, yep, the robots'll win. But be warned: They are frustrated by not being humans like Pinocchio crying to be a real boy, and ultimately, they'll kill us all. So many are so willfully, unabashedly ignorant yet eager to dictate; so many are against fundamental morality in favor of just agnostic science; so many adoptive of Orwellian language; so many prizing ease and convenience over competence...we are RIPE for takeover—*why not robots?* Or, maybe worse, a small cabal of globalists in control of AI?

Years ago, *Tonight Show* host Jack Paar somehow had three cougar or leopard cubs living in his house, loose, as pets. He filmed them and showed it on TV: big cats playing like kittens. SO CUTE! One day he came home to find they had ripped the furniture to shreds, trashed the kitchen, ruined the house, and were lying in wait, to eat him. *NOT cute*. EVERYTHING THAT MAY ULTMATELY EAT US LOOKS "CUTE" AT THE START. Oh, look, Marxist professors and kids at college having anti-American, anti-Semitic, immoral ideas, going "woke"—isn't that funny? Oh, look, some moving into Corporate and putting men with beards in dresses in lipstick ads? Isn't that cute? Oh, look, self-driving cars and little robots—aren't they cute? Wasn't "social media" friendly 'n fun at the start?

AI UPDATE
(Maybe We "Have Nothing to Fear but Fear Itself")

(Date: Early 2024)

The first lawsuit against AI (Microsoft, Open AI, ChatGPT) for massive copyright infringement was filed by the *New York Times* in December 2023. If you are an author or publisher or you use a lot of content in your marketing or/and are concerned about being sued under a new standard, that might be established in this case—you should familiarize yourself with it and follow it. It's deep pockets vs. deep pockets, so it may reach the Supreme Court. It may also influence congressional legislation. I personally think it's doomed to fail. It seems to me it would make the acts of (a) speed reading a lot of material already published on a subject—say "narcissism"—then (b) writing articles or a book in your own words based on your amassed and comingled research, (c) without citing every individual thing you read (heard, watched) illegal. This is what most of us do, so this case may come down to: Does size matter? Should the *NYT* get any kind of settlement or favorable verdict; it could affect everybody using AI to create content—did the emails you wrote selling your juicer steal from the *NYT*? Or the estate of Jack LaLanne? Regardless of the outcome of this litigation, it raises bigger issues.

Same day I was writing this (myself!), I ran across an old, full-page magazine ad by Motorola headlined "Why Television Is So Good for Your Children." The ad shows a couple of young kids sitting on the floor, in front of a big TV, mesmerized. The copy extolls the virtues of this new technology that brings the whole world into your living room. It even advises sitting very close, to avoid eyestrain. WE EASILY FORGET that NO new tech has EVER lived up to its early concepts and promises. Almost all new tech has morphed into something very different from what was originally expected. NONE entirely free of adverse side effects.

This is from Sam Altman, CEO of OpenAI: "Even though in the long term, I think a lot of the hype about AI is warranted, the short-term hype is pretty disconnected from the current reality of the technology."

CHAPTER 13

Automating Direct Mail and Strategic Gifting

A Guide to Enhancing Customer Relationships

by Parthiv Shah

Mailboxpower.com is a software company that facilitates automation of personalization, printing, and shipping of printed letters, greeting cards, gifts, and Wow! boxes. While "personalization through mail-merge" is not a new concept, most businesses don't have in-house capacity to do personalized printing with personalized images, personalized QR codes, and almost no one has in-house capacity to do personalized gifts.

Every business wants to deliver an amazing experience in a box...but the logistics of doing everything to prepare a thoughtful gift box and do it for every prospect, every client, at every desired touch point is a near impossible task.

Mailboxpower.com tackles this challenge. Through an API gateway, I connected my HubSpot to Mailboxpower.com and created an automation in HubSpot to trigger a ZAP pushing a contact record from HubSpot to Zapier and Zapier to Mailboxpower.com and placed the record in an appropriate "list" in Mailboxpower .com. The "list" in Mailboxpower.com is connected to automatically enroll someone in the sequence of "packages" I've created, to be sent out by Mailboxpower.com.

The "sequence of packages" in Mailboxpower.com are called "automations." I have FOUR main automations built out:

- **Classic LTN (Long-Term Nurture)**
  - This is a monthly greeting card with a "happy holiday of the month" like New Year's, Valentine's Day, Father's Day, Mother's Day, etc.
  - This costs under $10 per year per person plus postage.
- **Premium LTN (Long-Term Nurture)**
  - This is a monthly "holiday of the month" where we invented holidays that are suitable for the companion gifts we are sending—"national popcorn day," "national pistachio day," "national hydration day," etc.
  - The cost of this is under $100 per year per person plus shipping. This includes cost of printing, personalization of greeting cards, and personalization of gifts.
- **VIP LTN (Long-Term Nurture)**
  - This is reserved for high-value clients. This is a THREE-gift sequence with a monthly greeting card sequence.
    - ◊ Gift 1: Charcuterie board personalized with their family name

◊ Gift 2: Six months later: Laser Engraved Wine Box and tools

◊ Gift 3: Personalized Water Bottle

This is accompanied by a monthly greeting card that gives you a recipe and shopping list of how you can use your Charcuterie board THIS MONTH.

Strategic Gifting Case History: The Story of Hole Hogz

Hole Hogz is a company specializing in hydro excavation—a modern, efficient method of digging using high-pressure water. Despite offering a cutting-edge service, they faced the challenge of breaking into a market dominated by traditional excavation methods. To tackle this challenge, Hole Hogz turned to strategic gifting, leveraging the power of personalized direct mail. They sent out a series of custom gifts to carefully selected potential clients. These weren't just any gifts; each item was thoughtfully chosen to resonate with their target audience—contractors who would benefit from their service.

The first package included a custom box, a personalized greeting card, and branded treats. It was a surprise, a token of appreciation, and most importantly, a conversation starter. The second and third mailings followed, each building on the impact of the last, with gifts like a personalized poker set and a customized journal, further solidifying the relationship.

These strategic gifts did more than just introduce Hole Hogz's services; they created a lasting impression, making each recipient feel valued and special. The result? An overwhelming response from the market, leading to significant business growth and the need to expand operations.

Dan Kennedy has a short list of advice about introducing yourself to a market or presenting yourself to a particular prospect:

- SHOW UP—a lot, often.
- SHOW UP "DIFFERENTLY" than all competitors.
- SHOW UP "BIGGER" than your prospects are used to from competitors and even everybody else communicating with them. If others have brochures, you need a book and a video. If they have that, you need a box of four books, two videos, and a toy truck (object relevant to your market or product).
- SHOW UP IN WAYS YOU CANNOT BE IGNORED.
- SHOW UP WITH BENEFICIAL VALUE AND/OR "FUN" while also advancing your cause and selling your products or services.

Personalization Adds Enormous Impact

Dale Carnegie became famous in the field of human relations with several very simple ideas, including that the sweetest sound anyone can ever hear is their own name.

Automation enables the personalization of gifts and messages on a large scale. Using customer data, businesses can tailor their gifts and notes to each recipient's preferences and history with the brand. This personal touch, applied efficiently to a broad audience, ensures each customer feels uniquely valued.

Efficient Management of Mailing Lists

Automated systems can manage and update mailing lists dynamically, ensuring accuracy and relevance. They can segment audiences based on various criteria, making it easier to target the right people with the right gifts, much like Hole Hogz identified and reached their ideal clients. List segmentation is uber-powerful.

Streamlining the Strategic Gifting Process

From selecting the right gift to packaging and shipping, automation streamlines the entire gifting process. This efficiency reduces the time and resources spent on manual tasks, allowing businesses to focus on the strategic aspects of their campaigns.

Tracking and Analytics

Automated direct mail systems offer robust tracking and analytics. Businesses can monitor the delivery and reception of their gifts, gather feedback, and measure the impact of their campaigns. This data is crucial for refining strategies and ensuring the highest return on investment.

Consistency and Timeliness

Automation ensures that gifts are sent out consistently and timely. For businesses like Hole Hogz, this meant maintaining a regular touch point with their customers, keeping the brand top-of-mind, and building anticipation for the next interaction.

By automating direct mail processes, businesses can elevate their strategic gifting campaigns, combining the personal touch of traditional gifting with the efficiency and precision of modern technology. In the next sections, we will delve deeper into integrating these automated strategies with broader marketing efforts and measuring success.

Complementing Digital Efforts

Automated direct mail should complement, not replace, digital marketing strategies. For instance, following up an email campaign with a tangible gift can reinforce the message and make a lasting impression. It's about creating a multi-channel approach that leverages the strengths of both digital and physical marketing.

Leveraging Data for Targeted Campaigns

Use customer data gathered from digital channels to inform your direct mail campaigns. This data-driven approach ensures that your gifts and messages are highly targeted and relevant, increasing their impact.

Enhancing Customer Journey

Consider the customer journey and how automated direct mail can enhance it at various stages. From welcoming new customers with a personalized gift to reengaging dormant clients, strategic gifting can be a powerful tool throughout the customer life cycle.

Integrated Campaign Planning

Plan your direct mail campaigns in conjunction with your digital marketing calendar. This integration ensures that your messaging is consistent and that each channel reinforces the others, creating a more powerful overall campaign.

Integrating automated direct mail with your broader marketing strategies creates a synergistic effect, enhancing the impact of each individual channel. By leveraging the strengths of both digital and physical marketing, you can create a comprehensive, multifaceted approach that resonates deeply with your customers.

Practical Steps for Implementation

To effectively implement strategic gifting and automated direct mail, follow these practical steps:

1. Identify Your Audience

Understand your target audience's preferences and needs. Segment your customer base to tailor your gifting strategy more effectively.

2. Select the Right Gifts

Choose gifts that resonate with your audience and reflect your brand values. The gifts should be thoughtful, useful, and memorable. Ideally, they stay on the desk, the bookshelf, in the home, get shown to others. Be imaginative. For my BRENT campaign, based on elephants never forgetting, I send little stuffed-animal elephants. For his NO B.S. brand, Dan once had aerosol cans of air freshener labeled as NO B.S. SPRAY for use anytime somebody was B.S.-ing you. These made people laugh—and they got displayed in recipients' offices. A lot of his Members and clients have also been gifted a bobblehead doll; Dan on the white "NO B.S." bull made famous on book covers and company media. Today, a great gift can even "go viral" on social media.

3. Utilize the Right Technology

Invest in a reliable direct mail automation system that can handle personalization, mailing list management, and tracking.

4. Craft a Compelling Message

Your message should be personal, reflect your brand's voice, and complement the gift. A well-crafted message enhances the perceived value of the gift.

5. Schedule and Plan Your Campaigns

Timing is crucial. Plan your campaigns around significant dates, customer milestones, or periods when a personal touch would have the most impact.

6. Monitor and Adjust

Continuously track the performance of your campaigns and be ready to adjust your strategy based on customer feedback and engagement metrics.

7. Maintain Data Privacy and Compliance

Ensure that your gifting strategies comply with data privacy laws and regulations. Respect your customers' preferences and data.

Implementing strategic gifting and automated direct mail requires careful planning, a deep understanding of your audience, and the right technology. By following these steps, you can create impactful, personalized campaigns that foster strong customer relationships and drive business growth.

Looking Ahead: Trends in Direct Mail and Strategic Gifting

Increased Personalization Through AI

Advancements in artificial intelligence will allow for even more personalized gifting experiences. AI can analyze customer data to suggest the most impactful gifts, enhancing the personal connection.

Integration with Digital Experiences

Expect to see a blend of physical and digital experiences. QR codes on gifts leading to digital content, augmented reality experiences linked with physical items, and online communities centered around the gifting campaigns are all examples of integration.

Sustainability in Gifting

There is a growing trend toward eco-friendly gifts and packaging. Sustainable gifting resonates with environmentally conscious consumers and reflects well on your brand's commitment to social responsibility.

Experiential and Immersive Gifts

Moving beyond tangible items, gifting experiences that create memorable moments will gain popularity. This could include exclusive event invitations, virtual reality experiences, or personalized online workshops.

Data-Driven Decision-Making

The use of data analytics in choosing, timing, and customizing gifts will become more sophisticated. This approach ensures that gifting strategies are effective and resonant with target audiences.

Staying abreast of these trends and adapting your strategies accordingly will ensure that your direct mail and strategic gifting efforts remain effective, relevant, and appreciated by your customers.

Dan Kennedy Says

The late, great publisher Marty Edelston, founder of Bottom Line Media, a friend, kept a Toy & Gift Closet at his office. For every visitor he had or sought a relationship with, he would, at some point, excuse himself, go to his closet, find a relevant and appropriate gift from the wildly diverse inventory, and gift it. These surprises were never forgotten. Marty said it was a giant, life-size version of finding your toy surprise in your box of Cracker Jacks—which now, incidentally, involves a scan code taking you to free games online, not a plastic animal or a baseball card. Sigh. Anyway, I've done my own version of Marty Edelston my entire career. My "VIPs" each get different, sometimes personalized but always personally relevant, thank-you and holiday gifts. People who visit my office leave with a souvenir gift. My newsletters have often been accompanied by little toys 'n gifts: a little pack of candy corn in October for Halloween, a small door hinge (tied to "little hinges swing big doors"). I am in B2B, not B2C, dealing with serious businesspeople about serious matters, but there's still room to surprise and *delight.*

The opportunities presented here to make all this easier, to make it personal to each individual yet also automate it, and to automate so you never forget or get too busy are fantastic. Why wouldn't you take advantage of them?

CHAPTER 14

Case History: How Using Artificial Intelligence to Create a Black Friday Challenge Funnel Hit a Sales Home Run

(Selling: How to Use AI to Do a Challenge Campaign)

from Jeff Hunter and Samuel Young

AI expert and entrepreneur Jeff J. Hunter had his reservations about doing a Black Friday promotion. Between running his AI training business and managing high-ticket consulting contracts already generating over $300K annually, he felt too busy to orchestrate a sales campaign himself. But then fellow marketing expert Samuel Young, one of Jeff's past students, reached out with a bold idea. He asked Jeff, "Got any plans for Black Friday?" Jeff said no, and Samuel saw an opportunity. He told Jeff, "I have a plan for how we can use AI to fully automate and optimize an entire Black Friday promo focused on your AI program. And I think if we work fast, we can hit $50K in sales. In one day."

Intrigued, Jeff agreed to move forward and brought Samuel onto the project on November 4. They had less than a month before Black Friday. Together, Jeff and Samuel combined their expertise across marketing, technology, and AI to execute a cutting-edge campaign. While ambitious in its scope and timeline, they knew leveraging AI to handle time-intensive processes like copywriting, design, and video production could enable their speed and efficiency.

Incredibly, just 25 days later, Samuel's plan worked better than envisioned. Through relentless AI automation to create assets and drive exposure combined with carefully crafted scarcity and urgency elements, they generated over $50K in sales before Black Friday even arrived. This chapter will document Jeff and Samuel's approach so others can adapt these techniques of rapid AI automation, growth hacking, and converged media to drive similar innovations and results.

It Began with Lead Generation

Jeff and Samuel knew generating high-quality leads would be key for their promotion's success. To kick things off, they decided to leverage AI to create a Lead Magnet that would provide immense value to their target audience of marketers.

They used Claude AI to generate the concept based on this prompt: **"Provide me with 5 Lead Magnet Ideas for the AI Persona Method."** Claude suggested several options, with the one that resonated being "100 AI-Powered Marketing Prompts," which aligned perfectly with their promotion's focus. However, rather than just running with Claude's initial phrasing for the prompts, Jeff and Samuel refined them extensively by hand to ensure maximum quality. This hybrid approach enabled them to tap into AI for ideation while still delivering a polished, human-crafted marketing asset. Once finalized, the prompts were placed behind an opt-in form as a free downloadable resource.

To drive traffic to this lead capture landing page, Claude AI was once again leveraged: **"Create an opt-in page for my Lead Magnet below. Lead Magnet: '100 Marketing Prompts.' Do not use the word 'Lead Magnet' in the copy; it should be referred to as prompts. Follow the StoryBrand Framework to create the copy."**

Next, Jeff and Samuel focused on further building out their list by enticing contacts to join a waiting list for their upcoming promotion. ChatGPT was used to generate email copy and social posts for this list-building outreach effort.

Within days, hundreds subscribed to their waiting list, clearly signaling high anticipation for their AI-powered Black Friday launch.

Using Savvy Market Research

Rather than just guessing what their audience might want from a Black Friday deal, Jeff and Samuel decided to go directly to the source and survey customers to determine their needs.

They used ChatGPT to generate a market research survey. Here are the survey questions:

1. *What are your top 3 most time-consuming tasks in your business?*
2. *What are your top 3 biggest expenses in your business?*
3. *What are 1–3 things that would help you grow your business that you do NOT have time for right now?*
4. *What are your top 3 questions about AI?*
5. *What are the top 3 things you want to learn about AI?*
6. *What challenges have you had with AI in the past, if any?*
7. *What is your total budget for the next 12 months for AI training if it helped you save a lot of time and money?*

The survey took an open-ended format, allowing respondents to explain in detail their own contexts and interests. Over 140 customers completed the survey, drawn in by the Lead Magnet incentive. While open-ended questions provided richness, analyzing all that qualitative data would be time-consuming for a human. Instead, Jeff and Samuel had ChatGPT take on the job by creating a dedicated data analysis persona.

"Analyze this research survey with data from my audience. Summarize the data and create a list of the top 3 most common answers in each category."

ChatGPT's persona reviewed all survey responses and quantified key patterns that emerged, capturing metrics like:

- The most frequently cited time-consuming tasks
- Top business expenses
- Budget ranges and willingness-to-pay levels

Typically, Jeff sold his training for around $29—$297 in special promotions—but Samuel felt the data told a different story this time. Customers longed for more robust, higher-priced solutions aligned to real pain points versus shallow discounts.

Convinced by the data, Samuel proposed disregarding conventional wisdom and instead creating a premium, AI-powered bundle packed with value and bonuses that could command top dollar. Seeing the figures himself, Jeff agreed to bet against his Black Friday pricing assumptions and aim higher, designing an irresistible high-end offer anchored at $997.

(How much money do you leave on the table by your own pricing assumptions, unverified by surveys, testing, or other research? Great reading on the entire subject of pricing is the book *NO B.S. PRICE STRATEGY*.)

The Birth of a "Best" Offer

Armed with quantitative data on their audience's needs and constraints, Jeff and Samuel were ready to formulate an irresistible offer. But rather than just relying on their own opinions, they wanted to tap into time-tested direct response principles. They decided to take the unconventional approach of actually training ChatGPT itself on the fundamentals of high-converting offers. Specifically, they used Alex Hormozi's acclaimed "$100M Offers" training, uploading transcripts covering multiple hours of video content from his course. This enabled ChatGPT to internalize Hormozi's proven frameworks on risk reversals, fast action bonuses, stacking value, and more. With ChatGPT now trained on structuring offers, Jeff and Samuel could leverage this training while incorporating their survey data and intended core program to script the perfect promotion.

But before finalizing, they also ran their draft offer through one more AI-based analysis. They uploaded lead offers from their competitors into ChatGPT and had it compare those to their proposed deal. ChatGPT provided additional suggestions for improvement opportunities, which they used to further refine specifics around deliverables and bonuses included.

Ultimately they settled on bundling Jeff's foundational AI Persona Method training with over a dozen hours of supplementary trainings focused on application in marketing, copywriting, social media growth, and lead conversion. These modules directly addressed the time-intensive activities and expensive optimizations their survey data indicated customers needed most help with. With the core offer in place, Jeff and Samuel added limited-time "fast action" bonuses that featured extra incentives for being first-in live 1-on-1 strategy consultations with an AI Consultant, and a bonus live lead generation workshop. They capped these to only the first 10 and 25 customers to further increase excitement and evoke FOMO with those still on the fence. This strategy played a key role in accelerating sales velocity out the gate.

The final step was to construct two upsells to increase the average cart value. Jeff and Samuel decided to add Jeff's AI Consulting Program for $1,497, and Jeff's high-ticket consulting for $9,000. The two upsells would dramatically increase profits during the sale.

Creating the Video Sales Letter (VSL)

Jeff and Samuel knew that having an exceptionally compelling video sales letter (VSL) would be critical for converting visitors into buyers. But writing an effective 10-minute script and filming high-quality footage would normally demand extensive effort.

Luckily, AI could automate the heavy lifting as long as they provided enough relevant training data. They leveraged

Figure 14.1: The Offer

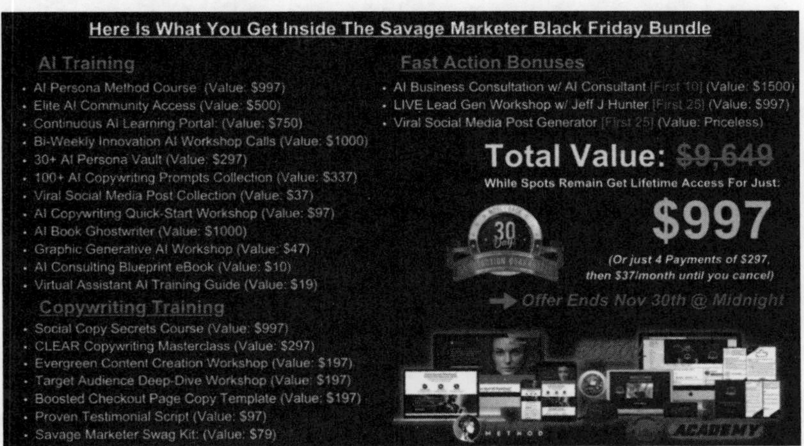

ChatGPT by constructing a dedicated VSL creator persona by training it on VSL Expert Kevin Anson's proven framework. They also uploaded Jeff's own materials to help ChatGPT capture his distinct personality, voice, and positioning. After sufficient priming, ChatGPT was ready to script an initial VSL outline. Jeff and Samuel reviewed it, had a few rounds of revisions, and arrived at the final story arc they felt flowed logically while embedding core offer details. With script locked in, recording, at first, followed an unorthodox approach. Rather than memorize lines or use cue cards, Jeff simply read off his computer screen so he could get the VSL recorded before the end of the day. The only catch? This meant that he was staring at the computer screen, and not directly into the camera lens while reading, but Descript and Captions AI tools offered easy fixes. Using footage of Jeff looking directly at the camera for a few seconds, these apps realistically overlaid his glance into the clips to have it appear his gaze held viewers' focus. Some simple lighting adjustments ensured an even blending for minimal uncanny valley effect.

Samuel also used Descript to effortlessly insert pickups, overdub lines, and tweak verbiage after the fact instead of refilming specific parts of the video.

Expert VSL video editor Spence Thompsett used Envato video elements to spice things up with slick graphics and editing effects in between cuts of Jeff speaking. By blending AI and human creativity into an impactful final cut, the 10-minute VSL went live just days before the Black Friday launch. Using AI tools to write the script and help with editing made creating an amazing VSL in a limited amount of time possible.

Generating Traffic

With their VSL and Sales Page complete, Jeff and Samuel focused next on driving traffic across media assets powered by AI. They repurposed sections of their VSL transcript into email sequences and social media content. However, rather than just reusing content as-is, they wanted AI to repurpose the content to be formatted for each marketing channel. By supplying ChatGPT with training data on proven copywriting approaches and summarizing key patterns from their market research, the AI could specially tailor emails. This allowed them to communicate their offer's applicability toward resolving recipients' precise pain points and challenges for further personalization. These emails Were aimed at progressively warming up the audience, showcasing solutions to pain points, highlighting benefits, and finally driving visitors to the sales page once live. Meanwhile Facebook and Instagram retargeting ads provided supplementary exposure. This cross-channel effort enabled reaching their audience at multiple touch points when visibility and conversion inclination were highest.

Here is an example of one of the emails we wrote with AI:

Hey [FIRSTNAME],

I know what you're thinking…

"AI sounds complicated."
"I don't have time to learn it."
"It doesn't work for my business."
"AI-generated content sucks and makes me sound like the Terminator."

Believe me, I've heard all the objections before. But here's the truth: My method makes AI simple and practical for any business. As a business owner myself, I know you don't have time for overly complex technology. My method breaks AI down into easy-to-implement steps tailored specifically to your business needs.

It's been tested now by a couple hundred business owners. You'll be able to start small by automating repetitive tasks like content creation. Then build up your AI capabilities as you become more confident.

I can't give you all the details of what I'm working on just yet… ButI can tell you that spots will be extremely limited! If this sounds like something you don't want to miss then here's the deal: Reply "AI" back to this email to join the waiting list today and be one of the first to learn about my crazy Black Friday offer. When you join right now, I will send you my 100+ Proven AI Marketing Prompts so you can start leveraging AI right away. Reply "AI" back to this email to join the waiting list today!

Jeff J. Hunter
Founder of the AI Persona Method

Optimization & Results

As new customers purchased the product, Jeff and Samuel kept fine-tuning aspects of the funnel itself, as well as audience targeting across media channels for maximum response.

For instance, they expanded their retargeting beyond just past site visitors to incorporate look-alike audiences that mirrored their existing customer profile. New visitors from these expanded pools proved highly qualified.

Just 25 days after originally conceiving this ambitious promotion, Jeff and Samuel leveraged the tireless productivity of AI to achieve their $50K launch goal before Black Friday even commenced. By the time the sale ended, over 50 additional customers were acquired through an AI-powered marketing process generating over $70K in total sales (this includes front-end and back-end sales). This case study clearly validates the power of an AI workforce in building a hyper-efficient marketing machine.

AI Tools Used in the Campaign

Jeff and Samuel assembled an ensemble of AI tools to streamline their marketing efforts. Here is a brief overview.

Claude AI: An advanced natural language model capable of generating long-form, humanlike content with minimal editing needed from raw outlines. Claude is better than ChatGPT for training on large data sets and copywriting.

ChatGPT: A popular conversational AI that can dynamically understand and respond to multifaceted prompts. Leveraged for ideation and creation across assets and analyses, ChatGPT is great for a fast workflow and handling multiple tasks.

Descript: An audio/video editing application with integrated AI speech to text allowing seamless voice overdubbing and manipulations to footage by editing the transcript.

Captions: A cost-effective video service that uses AI to auto-generate text captions and enables footage augmentation like eye/head movements.

Envato Elements: A stock media marketplace providing ready-made digital assets to incorporate in video and other projects. You can use the AI search tool to find exactly what you need for your video project.

DALL-E: A leading AI image generator that creates original icons, illustrations, and artworks based on text prompts invoking a desired style.

Facebook Ads Manager: Facebook's advertising platform features advanced targeting capabilities powered partly by AI, enabling fine-tuned and optimized ad delivery.

The Secret Weapon: AI Personas

While leveraging an ensemble of AI tools was essential to executing this ambitious campaign vision, none of it would have been possible without one secret weapon—AI Personas. An AI Persona is effectively a simulated "virtual employee" with specialized skills. Just as you would train a new human team member, by supplying AI assistants with customized contextual and experiential information you enable vastly superior performance on tasks aligned to that training.

Rather than taking a blank-slate generic AI approach, Jeff and Samuel crafted tailored AI Personas for the required workflow needs.

Copywriting Persona: Fed proven copywriting frameworks, audience data, promotions details, etc. to create marketing emails, social media posts, and funnel copy.

VSL Persona: Uploaded frameworks from proven industry sources to internalize high-level sales psychology and storytelling frameworks in order to create the script for the Video Sales Letter in record time.

Data Analyst Persona: Shared survey data so it could quickly analyze and recognize customer patterns and extract insights.

These specialized AI team members knew exactly how to get the desired output because of the frameworks and data sets they were trained on. These results would have never been possible without properly training the AI. So for anyone looking to maximize productivity via automation, the lesson is clear—invest time up front developing customized AI Personas to match project needs through deliberate training. By creating this capable digital workforce, Jeff and Samuel tapped into the true power of AI.

If you want to learn how you can build your own AI Personas and stay up to date on the latest AI strategies, check out the AI Persona method by going to www.elaunchers.com/persona.

Dan Kennedy Says

In The Phenomenon®—my system for accomplishing more in 10 months than in your previous 10 years—I talk about the SPEED-TO-MARKET FACTOR and the SPRINT FACTOR. Many, maybe most promising business ideas die *s-l-o-w* deaths. Purpose + Pressure = Productivity & High Performance. This explains how often an NFL team struggling to slowly move the ball suddenly looks like a different, far more potent team in their "2-Minute Drill." When you use a Sprint, like Jeff and Samuel did, you get that dynamic effect. Then, with Speed to Market, you create far greater interest in a marketplace. The Impending Event is far more exciting than something just happening over a long period of time or in the distant future. Putting these two Phenomenon® Triggers to work for yourself, in your business, is not easy, but it is easier now than just a handful of years ago, in part thanks to the technology discussed in this chapter.

CHAPTER 15

An Adventure with AI as a Copywriter

A Case History

by Parthiv Shah

There is a huge need for copywriting, particularly direct-response copywriting, and for skilled copywriters. There's a shortage. The top tier of expert freelance copywriters, which includes Dan Kennedy, is over-busy and quite expensive. A project placed with any of them needs to be of substantial opportunity and size. Many business owners need to function as their own copywriters, at least for much of their day-to-day needs. With all this in mind, I wondered if AI could speed up their learning curve to copywriting competency and confidence. Could it be good enough on its own for many purposes?

Dipti Kala, a retired squadron leader from the Indian Air Force, chose a path less traveled after her military service. While many of her peers transitioned to the corporate sector, Dipti decided to leverage her leadership skills in a different arena—business coaching. With a decade of experience leading a team of officers and technicians, Dipti found her passion in helping working mothers excel in entrepreneurship and corporate sectors. Her journey took a unique turn when she delved into the world of AI, particularly ChatGPT. Dipti recognized the potential of AI to enhance productivity. She studied. Did her research. She quickly realized why many users struggle when it comes to ChatGPT. She also came to grips with a fact about business coaching: You have to have clients to coach! Where do they come from? Self-marketing. What does that require? Prolific content writing and copywriting.

A Meeting of Minds

I first met Dipti in an online ClickFunnel Facebook group—our common background as soldiers in the Indian Air Force made for easy conversation. When Dipti learned that I had won the prestigious Two Comma Club award from ClickFunnels twice, and I learned about her vision in professional digital marketing, a fast friendship began.

When I learned about her pursuits with AI, I offered a challenge...

The Ultimate AI-Powered Copywriting Challenge

Can ChatGPT be tasked with writing sales copy and creating nurture content using proven techniques, strategies, and copy frameworks?

To complete this challenge, we had some questions to answer...

- What skills are necessary to use ChatGPT for copywriting? Can anyone with a decent understanding of language take this tool and create dynamic copy that has prospects lined up with credit cards in hand?
- Where will Dipti and her AI tools encounter challenges? We recognized the opportunity. Now we needed to experiment in order to see where this tool might fall short.
- How much time will Dipti need to spend, and how much time can AI save? At the end of the day, after prompts, edits, and rewrites, how many man-hours are saved?
- Will the copy and content be "market ready"? What tasks will Dipti and other humans need to perform after AI generates the copy? Will the copy be so good that it needs a single read-over, and then it's ready to go to a client?
- Ultimately, can AI make Dipti and I money? Can it save us enough time, free up enough resources, and produce a good enough result to be worth the effort?
- Can a trained ChatGPT user in India successfully create direct response copy for a law firm in Maryland? Will the writing be personalized enough not to sound like a thousand other pieces of marketing tripe that can be found on just as many bargain-basement websites?

We had high hopes.

Dipti and I saw ChatGPT's potential to grow her coaching practice and make her money. If the end product turned out to be good enough to be market-ready, she could save business owners a significant amount of money.

Copywriting done by a competent writer is not cheap. A single sales letter written by a copywriter capable of swaying hearts and minds and turning tire kickers into paying customers, then to raving fans, can cost tens of thousands of dollars and be worth every penny, and then some. Dipti knew that if she could train her coaching clients to use ChatGTP to create compelling, engaging, and market-ready content, she could grow rich while helping thousands of working moms earn a living.

I recognized that if this worked for Dipti, I could help other professionals use AI to write copy and create content so "anyone" could use eLauncher's proprietary systems. There would be another, additional answer to: "How do I get all the marketing copy and content copy I need written?" If our experiment worked, eLaunchers and its systems would no longer be reserved for high-end marketers with significant marketing budgets. We could monetize a "Jr." eLauncher's A.B.C.D. (Any Business Can Do It) in a dozen different ways.

Time was also a factor. When I commission my writers to create copy for ALL eLaunchers' assets we produce for a client, an experienced writer typically needs 90–180 days to deliver. These assets include website development, Shock & Awe packages, sales letters, long-term nurture sequences, and more. It's a mountain of work. I gave Dipti 45 days to finish the project. I wanted to know if ChatGPT could make the impossible possible. Can this technology, in the hands of an experienced user, allow me to deliver quality work to my clients in half the time it takes my best writer to create the same? Could small business owners use this approach on their own?

If this worked, Dipti would have an entire suite of copy for her consulting company—assets that would require a significant investment if implemented by eLaunchers. Creating these assets would not have been possible without AI, because learning the core concepts of Magnetic Marketing® and adopting eLaunchers' implementation protocol was not practical for Dipti. She is busy. She's a successful entrepreneur. A coach. A MOM with two toddlers. And a wife who supports the busy life of a working husband who is active in the corporate world. There are simply not enough Dipti hours to overcome this learning curve and complete this work in a reasonable time. We needed to see if AI and ChatGPT could accomplish this for her.

So, I gave her access to my Toolbox. I equipped Dipti with a treasure trove of assets to study and to model. These included:

- A masterpiece sales letter, Shock & Awe package, and other assets created by Dan Kennedy for my company, Dental Growth Machine.
- An expertly crafted Pixel Estate Website written by Everte Farnell, using Dan Kennedy's copy as a foundation.
- My million-dollar funnels—the same ones that won me two of the prestigious ClickFunnel's Two Comma Club awards.
- My entire Business Kamasutra campaign for marketing automation written by Everte Farnell, including…
  1. Lead Magnet Delivery Sequence.
  2. New Lead Sequence—used to persuade a prospect to book a call.
  3. Pre-Appointment Sequence—setting expectations.
  4. Post-appointment sequences designed to "react" to the prospect's next move.
     - Appointment no-show—aimed to persuade a prospect to reschedule.
     - Appointment no-sale—offers compelling reasons why they should reconsider.
     - SOLD—a welcome sequence that sets the stage for what they can expect next.
  5. Long-Term Nurture—52 emails used to "stay in touch" with prospects. Great for promoting top-of-the-mind awareness and establishing a business owner as a trusted authority.

You can visit www.dentalgrowthmachine.com and request a copy of the info kit. Read Dan Kennedy's masterpiece copy and study these exceptional copywriting assets for yourself.

Despite the rich resources, Dipti encountered challenges in defining her "WHO" (Ideal Customer Avatar), articulating her "OFFER," and establishing her "USP" and "VALUE PROPOSITION." In an effort to bridge the gap, I provided one-on-one consultations and study materials, including books like *Breakthrough Advertising*, *Magnetic Marketing*, and *Ultimate Sales Letter*.

The Unraveling Experiment

As the experiment unfolded, Dipti faced difficulties. The dream of a plug-and-play solution faded away. We were left with a wish list rather than a finished product. It quickly became evident that SIGNIFICANT human intervention was necessary. This led to a decision to quit the initial approach entirely.

Lessons Learned

In reflection, we recognized that AI is a powerful tool for speeding up the writing process, but acknowledged its limitations. AI can't replace the unique skills of a seasoned copywriter like Dan Kennedy. However, it can be a valuable ally, enabling efficient market research and content articulation.

Dipti and I will continue to work together. I'll teach her how to write, and she will teach me how to be more efficient with ChatGPT and Command Prompts. We will make each other more productive. Have a lot of fun. Become better friends. And add value to each other's business. What these two soldiers will never do is build a robot that can replace Dan Kennedy or another copywriter of his experience, skill, and *intuitive* capability. The conclusion is clear—AI is everyone's business, and learning how to use it is crucial. Still, it has its limits.

My new glasses can help me read, but won't help me understand French or Italian. AI is a marvelous tool that aids human abilities. But it won't revolutionize every aspect of human life. AI can't make Dipti an excellent copywriter. But it can help with market research and help her articulate what she wants to say to her audience. This way, a trained and competent copywriter can step in and write for Dipti's brand.

How NOT to Speak to the Machines

Below is some good advice on how to make the most out of the powerful asset that is ChatGPT. Dipti learned these by experience, through struggle. I understand them by having watched her work and hit bumps in the road.

AVOID vague and ambiguous prompts, such as, "Write something about the need for estate planning." Prompts like this result in overly broad or irrelevant content. You need to get more granular.

AVOID leading prompts like, "Explain why our law firm is the best!" This will generate content that seems pedantic and that is product-centered rather than client-centered.

AVOID overly complex instructions like telling Chat GTP: "Create a comprehensive technical report on understanding restorative dentistry." This will lead to confusion, a lack of focus, and a report filled with vague generalities. Divide your desired report into sections, chapters, and individual topics within the big umbrella subject, and assign them one at a time.

AVOID too many requests in a single prompt. You can't tell ChatGPT to write a blog post, create a video script, and design an infographic about a new product all at the same time. Think in terms of to-do items, not to-do lists.

BE PREPARED to refine and fine-tune what you get from ChatGPT or other AI, and to go back to it with additional work provoked by what was completed. If you'll review Preston Bates' Black Friday Promotion experience described in his chapter, you'll see this is exactly what he and his partner did.

At least for now, at the time I'm writing this, I view ChatGPT as a new, *additional* member of a copywriters' room; a new, *additional* member of a marketing team. I have in-house teams in the U.S. and in India, and, for content and copy, I work with a number of freelancers including Dan Kennedy, Everte Farnell, and Russell Martino, so I am accustomed to drawing on multiple contributors per project. We use "the mastermind concept," derived from Napoleon Hill's writings in *Think and Grow Rich* and in *Laws of Success*. Now, a ChatGPT marketing strategist and a ChatGPT writer can join the mastermind. Specific to copywriting by committee, a book Dan Kennedy recommends is *COPY LOGIC! The New Science of Producing Breakthrough Copy* by Michael Masterson and Mike Palmer.

Dan Kennedy Says

Clint Eastwood says, "A man's got to know his limitations." With the experiment Parthiv describes here, he discovered the importance of knowing a technology's limitations and, in this case, AI's limitations. Like Edison with his 10,000 tries at the light bulb, I can assure you Parthiv is back in the laboratory conducting more experiments with this.

I think, broadly, this comes down to a piece of advice I first heard from Jay Van Andel, co-founder of Amway. Jay said, "Delegate or stagnate." I have modified it to live it and to teach it as: Delegate or stagnate–*but don't abdicate*. Trying to totally escape work, particularly important work, or escape responsibility, is, itself, a lot of work!–and rarely proves successful. The terrific auto industry CEO Lee Iacocca said he *had to* frequently get down out of the office tower and onto the factory floor, observe, talk with the men, and get out to the dealers' showrooms where cars meet customers. Walt Disney often disguised himself to walk the parks as a customer. Tom Peters coined the term MBWA for this–Management By Walking Around. I don't believe in totally removing yourself from any functions of your company as long as it's yours. I think you stick your nose in everywhere. There's a formula: Responsibility = Control, Control = Responsibility. The more responsibility you abdicate, the less control you have. It's proportional.

I know a number of top copywriters (and prolific, successful novelists) who use "subwriters." A few have writing factories. More just have small teams. The "name" copywriter tweaks, fixes, polishes the drafts completed by their "subs." Personally, I've tried this once, didn't like it, and have never done it since. I have occasionally delegated research, interviewing testimonials, and other time-consuming work required *before* writing. But even that I prefer doing because it can yield surprises. Useful answers to questions I didn't know to ask. For the copywriters who do use "subs" or the team approach, I can certainly see how ChatGPT can be added. Maybe even, soon, replace some of those human workers. But no top copywriter will ever fully abdicate these jobs. They may delegate the scientific and mechanical parts, but they'll have to control the organization at the beginning, add the artistry along the way, and polish then approve the final work.

CHAPTER 16

AI Tool List for Sales and Marketing Tasks

by Parthiv Shah

Here is a catalog of some of the most interesting AI-driven sales and marketing tools, as of this writing. (Remember, this is a book locked in the time it was written. See Page xvii for access to updates.) These are tools you may wish to select from and experiment with, to meet your particular needs.

Automation Tools: ChatGPT and Perplexity—Conversations with a Twist

As we explore the landscape of automation, **ChatGPT** emerges as more than just a chatbot—it's a versatile assistant seamlessly handling customer service, marketing inquiries, and educational interactions. Developed by OpenAI, ChatGPT engages in conversations, answers questions, and flexes its creative muscles to generate content.

Perplexity steps into the limelight as an intelligent search engine disguised as a chatbot. Its role is to enable users to ask questions naturally, responding in real time with context-aware answers. It's not just about information retrieval; it's about creating interactive customer support experiences.

Predictive Analytics in Marketing: Crafting Personalized Experiences with ElevenLabs

In the realm of predictive analytics, **ElevenLabs** takes center stage. An AI-powered marketing tool designed to create tailored content for customers, it dives into the intricacies of customer interests, suggesting content likely to captivate them, and optimizes marketing campaigns. It crafts personalized marketing campaigns, transcending generic approaches.

Creative Content Generation: Midjourney, Jasper, and Writesonic—Unleashing Creativity

Midjourney emerges as an AI text-to-image tool, transforming textual descriptions into visually stunning images. Midjourney is your artistic companion, turning your words into captivating visuals. It breaks down barriers, making visual content creation accessible to everyone, regardless of design skills.

Jasper, a powerful AI writing assistant, steps onto the scene. From blog posts to social media content, Jasper crafts high-quality, engaging material in minutes.

Writesonic is another AI writing assistant with a versatile tool kit. From generating text to translating languages, Writesonic adapts to various linguistic needs. It's not just about multilingual content creation; it's about assisting in creative writing tasks.

Transcription and Editing: Otter.ai and Descript—Amplifying Productivity

Otter.ai emerges as a robust AI-powered transcription tool, transcribing audio and video files into text in real time.

Descript steps into the limelight as a text and audio editing app using AI to streamline the editing process.

CHAPTER 17

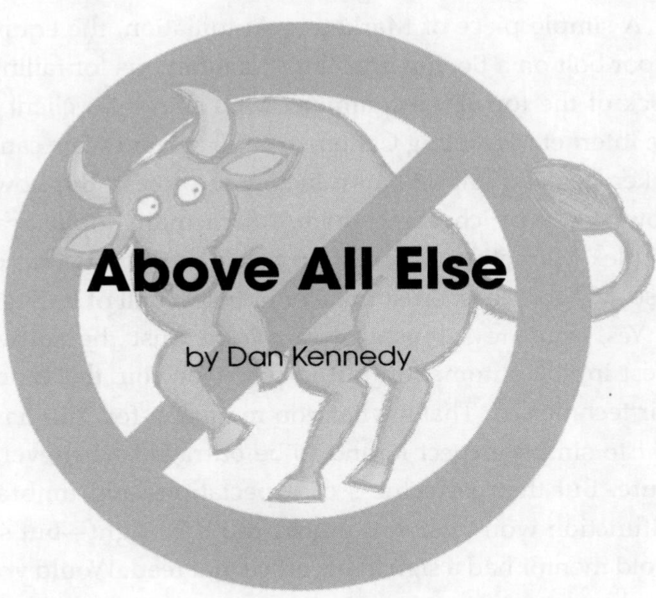

Above All Else

by Dan Kennedy

There is one thing more important than any other thing that can be said about Marketing Automation. It is an ancient management axiom: *Don't expect what you don't inspect—constantly.*

As I was working on this chapter, all Boeing 737s were grounded for inspection or, more accurately, re-inspection after the entire door blew off one of them at 16,000 feet. No passengers injured. But there could have been. It turns out not only were there loose bolts on that plane. Inspectors found loose bolts all over all the planes. Unfortunately, you can't personally inspect the innards of the Boeing-made plane you get on, expecting its bolts to be tight, its door to stay on for the entire flight. But you can personally inspect your own Marketing Automation. You or a highly, highly trusted person put in charge of doing so. Plus, maybe, "mystery shoppers." Every day. Maybe every hour. President Reagan had it right: "Trust, but verify." Verify, a lot.

A simple piece of Marketing Automation, the equivalent of a door bolt on a Boeing, the "link" is infamous for failing. In one check of the top 100 e-commerce web sites by a client of mine, The Internet Marketing Center, over 70 percent were caught with broken or dead links. "Not mine!" you protest. But how do you know? Only by checking them. Much more sophisticated and complex Marketing Automation has a lot of bolts that might be loose. My advice is: Trust none of it. Inspect all of it. Every day.

Yes, you are *supposed to be able to* trust the software you invest in, the automation put in place for you, the vendors and their technicians. That's what you're paying for. You have every right to simply expect it, and to be outraged whenever there is failure. But that correctness of expectations and umbrage with malfunction won't get you much. You'll be right—but so what? An old mentor had a sign in his office that read: Would you rather be right or rich?

I have a 50+ year habit of inspecting what I expect, often to others' annoyance. Moons ago, when I had an office and receptionists, and went to lunch, sometime during my lunch I went to the pay phone and called my own office just to check that the phone was being answered. First thing every morning, 7 days a week, from home, I faxed the office's fax machine and sometimes caught malfunction, then calling a staff-person at their home and telling them to haul ass to the office. Money came in through that machine. All of that is elementary compared to everything needing to be inspected with Marketing Automation. It's still valid regarding people, though, and is addressed in my book *NO B.S. RUTHLESS MANAGEMENT OF PEOPLE AND PROFITS, 2ND EDITION*.

All your automation MUST work 24/7/365. Customers or potential customers frustrated by it not working go elsewhere. Judge you as inept and untrustworthy. Money lost is unknowable, but ask people what they do when they bump up against automation that frustrates them. They'll tell you. Often with very colorful language. I sat once watching a friend try to order a whiz-bang, high-tech, very expensive office chair for me at its manufacturer's website, on Christmas Day, as her gift to me. Everything was fine until it came to checking out. She fought it for almost 30 minutes before saying, "The hell with them." Then, worse, she told the story, naming the company, in newsletters and speeches for a few years—as did I. No point naming the company now. They went bankrupt.

Just as no human employee is free of the price of supervision, no automation is either.

What Could Go Wrong? A Cautionary Tale.

From Christopher Steiner's great book *AUTOMATE THIS: How Algorithms Came to Rule the World* comes this tale of woe, slightly paraphrased and abbreviated…

In early April 2001, Michael Eisen, an evolutionary biologist, logged onto Amazon to buy a book, *The Making of a Fly* by Peter Lawrence, about the development from a single-celled egg to a fly, for his lab. The 1992 book, out of print, remains popular with academics and grad students. He expected to pay $35 or $40 for a used copy, as he had in the past. But on this day there were only two Amazon sellers offering new, unused copies—for $1,730,045 and $2,198,177. $1.7 million. $2.1 million. He assumed errors and checked back the next day to find that the prices had gone up! The price escalation continued day by day for two weeks, peaking on April 18, at $23 million. Plus $3.50 for shipping and handling. Finally, the next day, the price fell to $106. What had happened was that the ***unsupervised*** algorithms that priced books for the sellers got into a price war. One had been programmed to price the book just slightly higher than the competitor's price. The second "robot" ***automatically*** increased its price to be closer to that of the highest seller, prompting the first one to respond with another price hike. And so it went until the books were as expensive as Manhattan penthouses. Were there no adults in the room? No. Things didn't return to normal for 2 weeks until a human finally noticed, stepped in, and manually overrode the system.

If this were but one funny story, who cares?

But back on May 6, 2010, the Dow dropped 998.5 points in hours, the largest fall in history. Nearly $1 trillion of wealth disappeared. You can research the rest of the story if you like. It has a happy ending, sort of. There have been many more similar "incidents" since, with more to come, because unsupervised algorithms, left to their own devices, can do very strange things. They normally behave as they're designed, until they don't, or until some latent design flaw is suddenly exposed. Today's algorithms intersect with AI, actually increasing the possibilities for mistakes, from small to epic. There are also risks of outsiders' hacking and platforms' manipulations. The suppressed Hunter Biden Laptop Story could just as easily be your e-commerce site or your blog. You might think it's there, but where *did* all the traffic go? Pro-digital marketers know full well that Google, YouTube, Facebook, and the others *can't* be trusted. I'll argue that NOTHING having to do with tech or online spaces can be trusted.

In the must-read book *Get Rich or LIE Trying: Ambition and Deceit in the New Influencer Economy*, Symeon Brown puts it this way:

> *"YouTubers, Instagrammers, and TikTok creators are all at the whim of the rules, payment models and algorithms of platforms they do not own, cannot control and upon which they are utterly reliant. They do not get a vote on how enagement should be counted or how easy (or difficult) it is to find their*

pages, and how often they find themselves disadvantaged by arbitrary adjustments. Creators on YouTube have seen their income change every time the algorithm does."

There's more. A lot more. To be aware of, understand, and strive to safeguard your business from. Whatever you automate can't be thought of as "done."

I feel fortunate at this stage of my life that I don't have to use any of it personally, and I don't. Not a bit of it. I haven't even seen Amazon or Facebook. But all but a couple of my clients use a Pandora's box of it and need Marketing Automation, so for and with them, I must deal with it. It has been very beneficial to be too old to have grown up with it glued to my paw while still in the crib, and to distrust it as totally and fiercely as I do. I've been wary and vigilant when, left to their own devices, clients wouldn't be. Parthiv is, of course, much, much, much more conversant with it all than I am. To the extent that I am a modern Luddite by choice, he is the opposite, immersed in tech. Before writing this final chapter for this book, I asked him: If you could only tell people one thing about tech and automation and its brave new world, what would it be? Without hesitation, he provided the old management axiom from the 1930s: Do not expect what you fail to inspect.

In 1954, the year of my birth, scientists at General Electric predicted that, by 1964, your television would be so thin it would hang on your wall like a picture and be able to change channels by voice command. We got there, late. It's new in these and other ways. Yet it's NOT new as a business. Content must still be created, most of it in familiar formats. Viewers must still be wooed. Advertising must still be delivered—as the streaming services that started out advertising free have discovered. Hit programs like *America's Got Talent* and *American Idol* are re-imaginings of talent competition shows from the dawn of TV itself. The hit streaming series *Yellowstone* has *nothing* new in it. Every plot twist, every character comes re-birthed from popular Westerns of the '60s forward. And so on. Everything old is new again. Everything new is old repackaged.

So, as this book has presented, technology has brought you an incredible array of Marketing Automation tools and opportunities. But what is required to capitalize on them is *not* new at all. Attractive and persuasive advertising messages and content must still be created. Marketing must still be implemented. Customers still wooed, romanced, married in ongoing relationships. Data gathered and used. Everything in Claude Hopkins' revered, classic book *Scientific Advertising*, written more than 80 years ago, still applies, even as tech allows it to be more scientific, and automation allows better capitalization. This is what keeps this gaining dinosaur—me—relevant!

You will succeed by bridging "old" with "new"—not abandoning the old for the new.

Other Books by the Author

Other Books in The NO B.S. Series

No B.S. Time Management for Entrepreneurs, Fourth Edition with Ben Glass (2024)

No B.S. Direct Marketing, Fourth Edition with Darcy Juarez and Marty Fort (2024)

No B.S. Guide to Succeeding in Business by Breaking All the Rules (2024)

The Best of No B.S. (2022)

No B.S. Guide to Direct Response Social Media Marketing, Second Edition, with Kim Walsh Phillips, (2020)

No B.S. Marketing to the Affluent, Third Edition (2019)

No B.S. Guide to Powerful Presentations with Dustin Mathews (2017)

No B.S. Guide to Maximum Referrals and Customer Retention with Shaun Buck (2016)

No B.S. Ruthless Management of People and Profits, Second Edition (2014)

No B.S. Guide to Brand-Building by Direct Response (2014)

No B.S. Trust Based Marketing with Matt Zagula (2012)

No B.S. Grassroots Marketing with Jeff Slutsky (2012)

No B.S. Guide to Marketing to Leading Edge Boomers & Seniors with Chip Kessler (2012)

No B.S. Price Strategy with Jason Marrs (2011)

No B.S. Business Success in the New Economy (2010)

No B.S. Sales Success in the New Economy (2010)

No B.S. Wealth Attraction in the New Economy (2010)

Forthcoming NO B.S. Books

No B.S. Guide to Growing a Business to Sell for Top Dollar with David Melrose (2025)

Other Books of Note by Dan S. Kennedy

Almost Alchemy: Make Any Business of Any Size Produce More with Fewer and Less (Forbes Books, 2019)

My Unfinished Business: Autobiographical Essays (Advantage, 2009)

The New Psycho-Cybernetics with Dr. Maxwell Maltz (Prentice-Hall Press, 2002)

AUDIO BOOKS are available at Audible.com.

Index